HILBERT'S PROGRAM

SYNTHESE LIBRARY

STUDIES IN EPISTEMOLOGY,

LOGIC, METHODOLOGY, AND PHILOSOPHY OF SCIENCE

Managing Editor:

JAAKKO HINTIKKA, *Florida State University, Tallahassee*

Editors:

DONALD DAVIDSON, *University of California, Berkeley*
GABRIËL NUCHELMANS, *University of Leyden*
WESLEY C. SALMON, *University of Pittsburgh*

VOLUME 182

MICHAEL DETLEFSEN

Department of Philosophy, University of Notre Dame

HILBERT'S PROGRAM

An Essay on Mathematical Instrumentalism

D. REIDEL PUBLISHING COMPANY

A MEMBER OF THE KLUWER ACADEMIC PUBLISHERS GROUP

DORDRECHT / BOSTON / LANCASTER / TOKYO

Library of Congress Cataloging-in-Publication Data

Detlefsen, Michael, 1948–
   Hilbert's program.

   Bibliography: p.
   Includes index.
   1. Mathematics—Philosophy. 2. Hilbert, David,
1862–1943. I. Title.
QA9.2.D48   1986   510′.1   86–6460
ISBN 90–277–2151–3

Published by D. Reidel Publishing Company,
P.O. Box 17, 3300 AA Dordrecht, Holland.

Sold and distributed in the U.S.A. and Canada
by Kluwer Academic Publishers
190 Old Derby Street, Hingham, MA 02043, U.S.A.

In all other countries, sold and distributed
by Kluwer Academic Publishers Group,
P.O. Box 322, 3300 AH Dordrecht, Holland.

Printed in The Netherlands.

To Martha, Hans, Anna and Sara

TABLE OF CONTENTS

PREFACE   ix

ACKNOWLEDGEMENTS   xiii

CHAPTER I: THE PHILOSOPHICAL FUNDAMENTALS
OF HILBERT'S PROGRAM   1

1. Introduction   1
2. Hilbertian Instrumentalism   3
3. Hilbert and Quine: Frege's Problem Revisited   24
4. Concluding Summary   34
5. Notes   36

CHAPTER II: A CLOSER LOOK AT THE PROBLEMS   45

1. Introduction   45
2. The Status of Induction   46
3. Poincaré's Problem   59
4. The Dilution Problem   62
5. Notes   73

CHAPTER III: THE GÖDELIAN CHALLENGE   77

1. Introduction   77
2. The Standard Argument   78
3. The Stability Problem   80
4. Strict Instrumentalism   83
5. The Convergence Problem and the Problem of Strict
   Instrumentalism   85
6. Conclusion   90
7. Notes   91

CHAPTER IV: THE STABILITY PROBLEM   93
1. Introduction   93
2. The Standard Choice of $\mathscr{C}$ 94
3. Arithmetization   97
4. Mostowski's Proposal   101
5. The Kreisel—Takeuti Proposal   113
6. The Classical Proposal?   124
7. Conclusion   129
8. Notes   129

CHAPTER V: THE CONVERGENCE PROBLEM AND THE PROBLEM OF STRICT INSTRUMENTALISM   142

1. Introduction   142
2. Localization   145
3. The Problem of Strict Instrumentalism   148
4. The Convergence Problem   152
5. Conclusion   155
6. Notes   157

APPENDIX: HILBERT'S PROGRAM AND THE FIRST THEOREM   161

REFERENCES   179

INDEX   183

PREFACE

Hilbert's Program was founded on a concern for the phenomenon of paradox in mathematics. To Hilbert, the paradoxes, which are at once both absurd and irresistible, revealed a deep philosophical truth: namely, that there is a discrepancy between the laws according to which the mind of *homo mathematicus* works, and the laws governing objective mathematical fact. Mathematical epistemology is, therefore, to be seen as a struggle between a mind that naturally works in one way and a reality that works in another. Knowledge occurs when the two cooperate.

Conceived in this way, there are two basic alternatives for mathematical epistemology: a skeptical position which maintains either that mind and reality seldom or never come to agreement, or that we have no very reliable way of telling when they do; and a non-skeptical position which holds that there is significant agreement between mind and reality, and that their potential discrepancies can be detected, avoided, and thus kept in check. Of these two, Hilbert clearly embraced the latter, and proposed a program designed to vindicate the epistemological riches represented by our natural, if non-literal, ways of thinking. Brouwer, on the other hand, opted for a position closer (in Hilbert's opinion) to that of the skeptic. Having decided that epistemological purity could come only through sacrifice, he turned his back on his classical heritage to accept a higher calling. And if that meant living a life of epistemological poverty in the eyes of the world, then so be it; poverty would find its consolation in purity.

To Hilbert, this was hollow piety. *Homo mathematicus* sometimes goes astray, but he is not born in sin. Therefore, the answer is not for him to deny his nature, but rather to take precautions to see that he does not succumb to the false beauty of paradox. This way, he can live a life not only of holiness, but also of abundance;

and such a life is surely preferable to that of the needless self-denial preached by the intuitionists. In this spirit, Hilbert boldly set forth his program to vindicate the patterns of reasoning to which *homo mathematicus'* cognitive nature predisposes him.

This program had two parts; one descriptive, the other justificatory. The aim of the former was to produce the sort of description of our natural ways of mathematical reasoning that makes precise evaluation of that reasoning possible. For Hilbert, this amounted to formalizing mathematical thought in the manner so ably illustrated by the work of Russell and Whitehead. Thus, he was led to say that

The formula game that Brouwer so deprecates has, besides its mathematical value, an important general philosophical significance. For this formula game is carried out according to certain definite rules, in which the *technique of our thinking* is expressed. These rules form a closed system that can be discovered and definitively stated. The fundamental idea of my proof theory is none other than to describe the activity of our understanding, to make a protocol of the rules according to which our thinking actually proceeds. Thinking, it so happens, parallels speaking and writing: we form statements and place them one behind another. If any totality of observations and phenomena deserves to be made the object of a serious and thorough investigation, it is this one . . . (Hilbert [1927], p. 475)

The aim of the justificatory part of Hilbert's Program was to produce a finitary proof of the reliability of our natural modes of mathematical reasoning. And, as is well known, it is this part of the program that became the target both of a massive technical assault by Gödel's Theorems, and of a sustained philosophical attack based on anti-Cartesian epistemological views. In our opinion, most of this criticism is mistaken. We take Hilbert's Program, as reconstructed here, to be a philosophically sophisticated and convincing defense of mathematical instrumentalism. And we also believe that it can withstand the usual technical criticisms based on Gödel's Theorems. These are the central themes of this essay, and the main reasons for writing it.

In Chapter I we set out the chief philosophical problems confronting Hilbert's instrumentalist outlook. These include (1) a problem of Frege's concerning the compatibility of treating a mathematical proof both as a purely formal entity and as a means of knowledge towards the proposition expressed by its conclusion, (2) a problem of Poincaré's concerning a possible vicious circularity in Hilbert's metamathematical use of induction, and (3) a problem concerning how the instrumentalist can be sure that in taking the epistemic shortcuts provided by his instrument, he does not dilute the epistemic quality of the product that results from its use.

In Chapter II we show how Hilbert's Program is equipped to deal with the first two problems, and how it is out of concern for the third problem that its finitism is best motivated. The rest of the main body of the text is given to a discussion of the third problem, which we call the Dilution Problem.

This problem is given urgency by Gödel's Second Theorem, which threatens its solvability by allegedly ruling out a finitary proof of the reliability of all but the weakest systems of mathematics. The argument underlying this allegation is set out in considerable detail in Chapter III, and its more serious inadequacies are exposed in Chapters IV and V. Finally, in the Appendix we contend with some of the supposedly anti-Hilbertian consequences of Gödel's First Theorem, dismissing them as unfounded.

The end result, we hope, is an essay which helps the philosopher of mathematics and the philosophically interested logician or mathematician to come to a better understanding and appreciation of Hilbert's Program and mathematical instrumentalism generally. If it goes even a little way toward counteracting the common misunderstandings of Gödel's Theorems and the crude charges of formalism that have sullied the literature on Hilbert's Program, we will consider it a success.

ACKNOWLEDGEMENTS

The writing of this book was begun in earnest during the 1981—
82 academic year while I was a Fulbright professor in the
philosophy department of the Filosofski Fakultet in Zadar,
Yugoslavia. I count it a privilege to have been there, and would
like to thank the members of the philosophy department for
having helped to make my stay so enjoyable and rewarding. A
particular debt of gratitude is owed to professors Heda Festini,
Nenad Mishchevich, and Arne Markusovich for many a pleasant
hour spent doing philosophy together.

While in Yugoslavia, I also had the opportunity to discuss my
work with groups of logicians and philosophers at several other
universities. And so I should like to offer my general thanks to
those participating in colloquia at the universities of Belgrade,
Ljubljana, Rijeka, and Zagreb, and at the Institute for Post-
Graduate Studies in Dubrovnik. More particular thanks is owed
those who engaged me outside the formal confines of the colloquia: especially, Virgilio Mushkardin, Vladimir Razhenj, and
Miljenko Stanich at Rijeka; Zlatko Klanac, Dean Rosenzweig,
Kajetan Sheper and Zvonimir Shikich at Zagreb; Kosta Doshen
and Aleksandar Kron at Belgrade; Matjazh Potrch, Valter Motaln,
Andrej Ule, and Tomo Pisanski at Ljubljana; and W. H. Newton-
Smith and Srdjan Lelas at Dubrovnik.

There are also many on this side of the Atlantic who have
helped me in one way or another. Among individuals, my greatest
debt is to Mike Resnik for having read and commented upon
nearly all parts of the manuscript. Also, it is he who first helped
me to understand clearly the nature of the fundamental disagreement between Frege and Hilbert. Others who read and offered
helpful comments on parts of the manuscript include Michael
Byrd, Aron Edidin, Hartry Field, Tim McCarthy, Tadashi Mino,

Fred Schmitt, Stewart Shapiro, and Steve Wagner. Finally, I owe a great deal to the writings of Hilary Putnam in the philosophy of mathematics generally, and, more specifically, to the work of Georg Kreisel, Dag Prawitz, William Tait, and Judson Webb on Hilbert's Program. My lack of agreement is by no means an adequate guage of my indebtedness to them.

In addition to the individuals mentioned, I am grateful to several institutions for their financial support of my work. Specifically, I must thank the National Endowment for the Humanities, the Fulbright-Hays Foundation and the University of Minnesota for grants and leaves of various sorts.

Finally, I am more than grateful to my wife Martha, and my children Hans, Anna, and Sara for their steadfast love and good-humored patience. I dedicate this work to them with love.

CHAPTER I

THE PHILOSOPHICAL FUNDAMENTALS OF
HILBERT'S PROGRAM

1. INTRODUCTION

In this chapter I shall attempt to set out Hilbert's Program in a way that is more revealing than previous treatments. Specificially, I shall try to improve upon preceding accounts of Hilbert's Porgram in each of the following respects: (1) in isolating and developing the distinctive form of mathematical instrumentalism which is basic to Hilbert's Program, (2) in defining the central problems which must be overcome in order to successfully defend any form of mathematical instrumentalism, (3) in showing how Hilbert's finitism can be related to his instrumentalism by natural and illuminating means, and (4) in identifying some important philosophical implications of the program which have hitherto not been brought to light.

The distinctiveness of Hilbertian instrumentalism has two bases. One is its superior sensitivity to the problems alluded to in (2) (in particular, to what we shall come to refer to as Frege's Problem and the Dilution Problem), and the other consists in its advocacy of only a *restricted* form of mathematical instrumentalism. Both of these features serve to distinguish it from other proposals of an instrumentalist bent; most notably that recently set forth by Hartry Field. Its responsiveness to Frege's Problem saves it from the dubious slide to nominalism which characterizes Field's argument, and its appreciation of the Dilution Problem helps us to see in a wholly new way why an instrumentalist might come to place a special value on finitary reasoning in metamathematics. Its restrictedness also reflects admirable subtlety. For it arises out of a consideration which seems to have escaped the notice of others: namely, that the metamathematical reasoning that must be appealed to in order to establish the reliability of the instrumental

1

methods may be seen as epistemologically equivalent to certain sorts of mathematical reasoning. Hilbert, at least, thought that this was so. And since, in order to solve Frege's Problem, he was obliged to treat the metamathematical reasoning in question as genuine, contentual reasoning, he was also obliged to treat the epistemologically equivalent mathematics in the same way. Hence the necessity to treat some mathematics as contentual rather than instrumental in character.

The restrictedness of Hilbertian instrumentalism also plays a role in determining its philosophical importance. As I shall treat of it, the philosophical interest of Hilbert's Program consists, in part, in what it has to say about the controversy between realism and nominalism in the philosophy of mathematics. Its chief novelty in this connection is the refinement which it gives to our under-standing of how an ontology and epistemology for mathematics might be determined; a refinement which defines a reasoned position lying somewhere between the realism of Frege and Quine and the nominalism of Field. Like Field, we shall challenge the Quinean doctrine that the *application* of a piece of mathematics (i.e., its use in deriving truths) exacts an ontological tribute from *that* piece of mathematics. We believe that there are other ways of accounting for the applicability of a given piece of mathematics than to treat it as literal truth. However, unlike Field, we are not sanguine about the prospect of rendering mathematics ontology-free. For ontological commitments of an epistemologically serious sort appear to re-emerge when, like Hilbert, we attempt to account for the applicability of mathematics by giving a meta-mathematical demonstration of its reliability.

We shall trace the path of this re-emergence in what follows, but for now we may summarize our findings (roughly) by saying that Hilbert's Program, as we shall develop it, leads to a new view of how the ontological commitments of mathematics are to be determined. Field, we believe, is wrong in his nominalist con-jecture regarding the ontology of mathematics. Quine's realist position, on the other hand, is equally unconvincing. In the end, our way of developing Hilbert's Program points the way to a modified realism (or instrumentalism). According to this view, the ontological commitments of mathematics are located not in those parts of mathematics which we use to acquire knowledge, but rather in those propositions which are used to establish the reliability of the mathematics thus used.

2. HILBERTIAN INSTRUMENTALISM

For present purposes, we shall take instrumentalism with regard to a given body T of (apparent) theorems and proofs to consist in the belief that the epistemic potency of T (i.e., the usefulness of items of T as devices for obtaining valuable epistemic attitudes toward genuine propositions of some sort) can be accounted for without treating the elements of T literally (i.e., as *genuine* propositions and proofs), but rather as "inference-tickets" of some sort. The precise mechanism by which an inference-ticket is supposed to work is a matter that we shall take up in more detail later. But for the time being let us say this much: a theorem (or pseudo-theorem) t functions as an inference-ticket with respect to a genuine proposition P when it is used to acquire an epistemic attitude (e.g., belief, justified belief, or knowledge) toward P, but *not* by dint of our adopting an epistemic attitude toward t itself; similarly, a proof (or pseudo-proof) p of a genuine proposition P functions as an inference-ticket when it is used to obtain an epistemic attitude with respect to P, but *not* by virtue of our having adopted an attitude of belief regarding the truth of its premises and the truth-preservingness of its inferences.

This, of course, does not so much tell us what an inference-ticket is as what it is not. But for our own limited purposes, we can satisfy our need for a positive understanding of the concept by restricting our attention to the particular type of inference-ticket that interested Hilbert.

Broadly speaking, Hilbert was intersted in computational or calculary or, to use his own term, "algebraic" instruments. According to this general conception, a computation or calculation or algebraic derivation of a genuine truth is to be sharply distinguished from a genuine proof of that truth. A genuine proof provides for the epistemic acquisition of its conclusion by exploiting the judgments of truth regarding its premises and the judgements of truth-preservingness regarding its several constituent inferences. A computation, on the other hand, provides for the epistemic acquisition of a proposition P to which it leads by exploiting a judgment regarding its (the computation's) formal character and a meta-computational judgement affirming the *reliability* of computations having that formal character as guides to truths of the type exemplified by P. In calling a computation truthful (reliable, sound) we are, therefore, neither asserting that its pre-terminal formulae are true (or even meaningful) nor that its inferences are truth-preserving. Rather we are saying only that its terminal formula expresses a genuine truth.

For Hilbert, then the apparent propositions and proofs of mathematics are to be divided into two groups: (a) those whose epistemic value derives from the evidentness of their content (the so-called *real* or *contentual* propositions and proofs), and (b) those whose epistemic value derives from the role that they play in some formal algebraic, or calculary scheme (the so-called *ideal* or *non-contentual* pseudo-propositions and pseudo-proofs). The contentual/non-contentual distinction drawn by Hilbert is thus one which distinguishes between an item whose epistemic utility is a function of its genuine meaningfulness and evidentness, and an item whose epistemic utility is a meta-theoretical function of its purely formal or "algebraic" properties. Consequently, the real/ideal distinction of Hilbert's Program emerges as a distinction marking the different epistemologies underlying two disparate sorts of items found in classical mathematics.

Perhaps the clearest expression in Hilbert's writings of this general point of view is the following.

How, then, do we come to the *ideal propositions*? It is a remarkable circumstance, and certainly a propitious and favorable one, that to enter the path that leads to them we need only continue in a natural and consistent way the development that the theory of the foundations of mathematics has already taken. Indeed, let us acknowledge that elementary mathematics already goes beyond the point of view of intuitive number theory. For the method of algebraic calculation with letters is not within the resources of contentual, intuitive number theory as we have hitherto conceived of it. This theory always uses formulas for communication only; letters stand for numerals, and the fact that two signs are identical is communicated by an equation. In algebra, on the other hand, we consider the expressions formed with letters to be independent objects in themselves, and the contentual propositions of number theory are formalized by means of them. Where we had propositions concerning numerals, we now have formulas, which themselves are concrete objects that in their turn are considered by our perceptual intuition, and the derivation of one formula from another in accordance with certain rules takes the place of the number-theoretic proof based on content.

Hence, as soon as we consider algebra, there is an increase in the number of finitary objects. Up to now these were only the numerals, such as 1, 11 . . . , 11111. They alone had been the objects of our contentual consideration. But in algebra mathematical practice already goes beyond that. Yes, even when a proposition, so long as it is combined with some indication as to its contentual interpretation, is still admissible from our finitist point of view, as, for example, the proposition that always

$$x + y = y + x,$$

where x and y stand for specific numerals, we yet do not select this form of communication but rather take the formula

$$a + b = b + a.$$

This is no longer an immediate communication of something contentual at all, but a certain formal object, which is related to the original finitary propositions

$$2 + 3 = 3 + 2$$

and

$$5 + 7 = 7 + 5$$

by the fact that, if we substitute numerals, 2, 3, 5 and 7, for a and b in that formula (that is, if we employ a proof procedure, albeit a very simple one), we obtain these finitary particular propositions. Thus we arrive at the conception that a, b, $=$, and $+$, as well as the entire formula

$$a + b = b + a,$$

do not mean anything in themselves, any more than numerals do. But from that formula we can indeed derive others; to these we ascribe a meaning, by treating them as communications of finitary propositions. If we generalize this conception, mathematics becomes an inventory of formulas — first, formulas to which contentual communications of finitary propositions [hence, in the main, numerical equations and inequalities] correspond and, second, further formulas that mean nothing in themselves and are the *ideal objects of our theory.* (Hilbert [1925], pp. 379-80. Here I have substituted '*x*' and '*y*' for the "black letter" characters of the text.)[1]

Evidently, an instrument's ultimate justification must be its usefulness in helping us to perform tasks whose performance we value. Thus, if Hilbert's instrumentalistic advocacy of the so-called *ideal* methods of proof in classical mathematics is to be defended, it must be shown that they assist us in acquiring something that we value. In order to give such an account, one must first identify what the goods are that we stand to gain by using the ideal method, and then show how it is that using the ideal method assists us in securing these.

Regarding the first concern, we can say that the goods in question are epistemically valuable attitudes borne toward propositions both of an empirical and a mathematical variety (where the mathematical propositions in questions are the so-called *real propositions*).[2] Given this description of its goals, Hilbertian instrumentalism is aptly characterized as a form of epistemic instrumentalism. However, since the epistemic goals may themselves be subordinate to other goals, we make no claim of ultimacy for this description of the goals of Hilbert's instrumentalism.

Constructing a satisfactory response to the second concern is, at least at first glance, more difficult. For one is tempted to try to vindicate the ideal method by attempting to show that certain true real propositions can, in principle, be proven (and, hence, epistemically acquired) only through the use of ideal methods. However, such an idea is not only mistaken, but contrary to the very strategy which Hilbert himself proposed for the vindication of the ideal method. All real truths can, in principle, be proven by real or contentual methods. This much follows from the constructivity of the intended notion of real truth.[3] Moreover, the means by which Hilbert planned to establish the reliability of the ideal method as a guide to real truth clearly presupposes that this is so. For his proof of reliability was supposed to proceed by showing how one could transform a given ideal proof of a real formula R into a real proof of R. Since real proofs may be assumed to be sound, this would show that ideal proofs of real propositions are also sound.[4] This is the essential idea underlying Hilbert's well-known epsilon-elimination strategy. So, Hilbert did not take the ideal method to be capable of proving real truths which are in principle unprovable by real or contentual methods. But then what *is* the value of the ideal method?

In Hilbert's view, what recommends the ideal method is its superior *efficiency*. Generally speaking, the basis for this efficiency has been taken to lie in the fact that ideal reasoning can serve as an abbreviation of contentual reasoning. It is this feature which von Neumann emphasized in his assessment of the virtues of the ideal method. After setting out the basic tasks facing Hilbert's Program, he described the point of trying to carry out the program in the following words.

To accomplish [these] tasks . . . would be to establish the validity of classical mathematics as a short-cut method of validating arithmetical statements whose elementary validation would be much too tedious. But since this is in fact the way we use mathematics, we would at the same time sufficiently establish the empirical validity of classical mathematics. (von Neumann [1931], p. 52)[5]

Hilbert, however, sought to offer a somewhat deeper account of the efficiency of ideal reasoning. There is, he maintained, a profound psychological tendency to reason in ways that, at least in their formal character, correspond to the principles of classical logic.

In the domain of finitary propositions . . . the logical relations that prevail are very imperspicuous, and this lack of perspicuity mounts unbearably if "all" and "there exists" occur combined or appear in nested propositions. In any case, those logical laws that man has always used since he began to think, the very ones that Aristotle taught, do not hold. Now one could attempt to determine the logical laws that are valid for the domain of finitary propositions; but this would not help us, since we just do not want to renounce the use of the simple laws of Aristotelian logic, and no one, though he speak with the tongues of angels, will keep people from negating arbitrary assertions, forming partial judgements, or using the principle of excluded middle. (Hilbert [1925], p. 379)

Taking the principle of excluded middle from the mathematician would be the same, say, as proscribing the telescope to the astronomer or to the boxer the use of his fists. To prohibit existence statements and the principle of excluded middle is tantamount to relinquishing the science of mathematics altogether. For, compared with the immense expanse of modern mathematics, what would the wretched remnants mean, the few isolated results, incomplete and unrelated, that the intuitionists have obtained without the use of the logical ε-axiom? (Hilbert [1927], p. 476)

Hilbert thus stressed the "simplifying" and "unifying" effects of applying the ideal method. These effects, presumably, are the basis of its epistemic utility. This has been the typical result of introducing ideal elements into our mathematical thinking and, according to Hilbert, it will also be the result of "closing" our thought under the classical forms of inference.

Just as $i = \sqrt{-1}$ was introduced so that the laws of algebra, those, for example, concerning the existence and number of the roots of an equation, could be preserved in their simplest form, just as ideal factors were introduced so that the simple laws of divisibility could be maintained even for algebraic integers (for example, we introduce an ideal common divisor for the numbers 2 and 1 $+ \sqrt{-5}$, while an actual one does not exist), so we must here adjoin the ideal propositions to the finitary ones in order to maintain the formally simple rules of ordinary Aristotelian logic. (Hilbert [1925], p. 379)

So, on this account, it is its ability to simplify which is the source of the epistemic efficiency of the ideal method. By simplifying our thought, the use of ideal elements unifies it. And in unifying thought, more of our cognitive or epistemic concerns are brought under the purview of a manageably complex method of epistemic acquisition.[6] In a nutshell, the use of ideal elements widens the scope of our methods of epistemic acquisition without complicating those methods to such an extent as would make them humanly unfeasible. Similar extension of scope using purely contentual methods would be impossible since it would require methods too complicated to be humanly practicable.

DIGRESSION. It may be that there is some sort of "Darwinistic" argument lurking behind these conjectures of Hilbert's. Our continued survival and improvement as a species depends upon our ability to be efficient gatherers of epistemic goods such as knowledge and justified belief. And our ability to be efficient gatherers of epistemic goods is enhanced by our tendency to reason in forms which are classical. More generally, our ability to acquire epistemic goods is enhanced by our having an "algebraic" or "calculary" representation of the world, which allows us to circumvent contentual reasoning which can be of an epistemically paralyzing length or complexity. And our preference for classical logical forms is to be seen as contributing to our epistemic efficiency in exactly this way: it "algebraically" unifies our thought along one of its most basic dimensions, namely its logic, and thus facilitates the development of an efficient algebra or calculus of thought.[7, 8]

Or it may be that Hilbert simply saw this predisposition toward classical reasoning as a quirk of human nature, and not as something which could be taken as a genetic advantage for creatures sharing the balance of our cognitive traits. But neither the psychological speculation itself nor the question of its possible grounding is of any great concern to us. This is so because the main philosophical insights afforded by our investigation of Hilbert's Program are valid regardless of whether his conjecture concerning the brevity and simplicity of ideal proof ultimately proves to be right (which a considerable body of prima facie evidence suggests it is[9]). *End Digression.*

But even if we grant that ideal proofs are shorter and simpler than their contentual counterparts, we still fall far short of what is required of a defense of Hilbertian instrumentalism. For their brevity and simplicity does not by itself imply that ideal proofs are of any use as devices for acquiring *genuine* epistemic goods. And, as we shall now see, this is the substance of a problem which Frege put to Hilbert.[10]

Frege's Problem is that of explaining how the construction of an ideal proof I for a real (contentual) proposition R could conceivably be of any help in the fixing of an epistemically valuable attitude toward R. Since I is not composed of genuine propositions, it cannot be seen as giving, in itself, a reason for R (which is what a genuine proof, in Frege's sense, does). And if I doesn't provide a reason for believing R, how is it that the construction of

I is relevant to the epistemic acquisition of *R* in a way that, say, doing the rumba or bowling isn't? This, in essence, is Frege's worry concerning ideal proofs.

For Frege an inference or proof is a sequence of judgements[11] where each member of the sequence, save the last, serves as an epistemic basis for some later member. As such, an inference or proof is to be taken as possessing epistemic significance since it is supposed to provide a means, both causal and evidential, through which an epistemic attitude toward its conclusion can be fixed. This view emerges quite clearly in the following remark.

. . . an inference does not consist of signs. We can only say that in the transition from one group of signs to a new group of signs, it may look now and then as though we are presented with an inference. An inference simply does not belong to the realm of signs; rather it is the pronouncement of a judgement made in accordance with logical laws on the basis of previously passed judgements. Each of the premises is a determinate thought recognized as true; and in the conclusion, too, a determinate thought is recognized as true. (Frege [1903], p. 82)

What Frege is objecting to is Hilbert's use of ideal or non-contentual derivations. Hilbert's instrumentalistic claim is that the ideal derivations are epistemically potent; Frege's Problem is that he doesn't see how they *could* be. Ideal derivations are not, after all, sequences of genuine judgements where the last is an *epistemic* transform of those judgments which come before it. Instead, they are merely sequences of formulae or "signs", where the last element of the chain is simply a formal or syntactic transform of those formulae which precede it. Fregean genuine proofs provide their conclusion with an epistemic pedigree; Hilbertian ideal proofs provide their terminal formulae with only a formal pedigree. Consequently, Frege did not see how the use of ideal proofs could be efficacious in the production of epistemic attitudes toward the thoughts expressed by their formal conclusions.

From the fact that the pseudo-axioms do not express thoughts it . . . follows that they cannot be premises of an inference-chain. Of course, one really cannot call propositions − groups of audible or visible signs − premises anyway, but only the thoughts expressed by them. Now in the case of the pseudo-axioms, there are no thoughts at all, and consequently no premises. Therefore when it appears that Mr. Hilbert nevertheless does use his axioms as premises of inferences and apparently bases proofs on them, these can be inferences and proofs in appearance only. (Frege [1903], p. 86)[12]

Frege's Problem can, perhaps, be made somewhat clearer by breaking it down into two sub-problems. One of these sub-problems is concerned with how the construction of ideal proofs can be efficacious in the fixation of even the *lower* forms of noesis (viz., belief)[13]; the other is concerned with how the construction of ideal proofs could ever lead to any *higher* forms of noesis (e.g., justified belief or knowledge), even if we grant that it might be used to produce noesis of the lower form.

In a genuine Fregean proof, noesis with respect to the conclusion is the causal product of the previous epistemic events of judgement making up the proof; i.e., it is a causal product of the previous judgements. However, in an ideal proof, there simply is no causal ancestry of judgements upon which noesis with respect to the conclusion *could* be based. But then how is it that the non-epistemic acts which figure in the construction of an ideal proof should prove to be capable of producing noesis regarding its conclusion?[14]

So, the first of Frege's sub-problems is that of explaining how *any* epistemically significant attitude could come about as the result of constructing an ideal proof when it doesn't appear as the product of other activity of seemingly no lesser degree of epistemic significance, such as bowling or dancing. And the second sub-problem is that of explaining how an attitude fixed in this way could possibly attain to the status of, say, justified belief or knowledge when beliefs fixed through the other activities mentioned presumably would *not* achieve such a status. In other words, it is the problem of explaining how it is that using the ideal method is an appropriate strategy for pursuing *truth*. Even if we were to succeed in isolating a mechanism through which *belief* could be brought about as the result of constructing ideal proofs, it would remain for us to show that this mechanism constitutes a reliable means of acquiring *true* beliefs. It is this latter task that is the concern of the second sub-problem.

It is to obtain a solution to Frege's Problem that the Hilbertian instrumentalist makes the ascent to metamathematics. Frege was right to stress the epistemic impotence of ideal proofs taken by themselves. However, he was dead wrong to think that this forms the basis for a telling criticism of Hilbert's instrumentalism. Ideal theorems and proofs *are* to be regarded as strings of marks; but they are not to be regarded *merely* as such. For in their use as computational or algebraic devices they are to be *evaluated*, (though not *interpreted*) metamathematically. And it is this metamathematical evaluation that confers epistemic potency on an ideal proof by incorporating it into a genuine, albeit metamathematical, proof of its conclusion. In very sketchy detail, here's how this is supposed to work.[15]

The evaluation of a given ideal proof I is composed of two judgements: the first is a judgement to the effect that I is an object having a certain syntactical character, and the second is a judgement to the effect that an object having that syntactical character has a terminal formula which expresses a true real proposition. Taken together, these two judgements amount to a genuine metamathematical proof of the real proposition expressed by I's terminal formula.

Accordingly, the Hilbertian instrumentalist does not contend that noesis with respect to a given real proposition is brought about purely as a result of the formal activity involved in constructing an ideal proof for it. Rather, he holds that such formal activity becomes noetically significant by being *contentually* assessed at the level of metamathematics. This contentual metamathematical assessment, constituting as it does a genuine proof of a real mathematical proposition, can then be used as an epistemic replacement for real *mathematical* proofs of that proposition without violating Frege's strictures.[16] Let me take just a moment to explain this.

As we have seen, the first of the two obstacles posed by Frege's Problem is that of explaining how the construction of an ideal proof for a real proposition might result in a causal basis for (the event of) taking an epistemic attitude toward its conclusion. In Frege's view, a genuine proof proceeds causally to noesis with respect to its conclusion on the strength of preceding noetic acts regarding its constituent premises and inferences: *judgements* regarding the truth of its premises and the truth-preservingness of its inferences form the causal basis for *judgement* regarding its conclusion.[17] Hence, if a computation is to be taken as an epistemically significant item, it must be treated as a genuine proof. In this way, the beliefs in the truth of its constitutent premises and the truth-preservingness of its constituent inferences combine to form a causal basis for belief in its conclusion. This is the Fregean conception of how computations gain epistemic significance.

On Hilbert's view, computation, of which ideal proof is but one variety, is to be treated differently. Like Frege, Hilbert believes that a computation takes on epistemic potency only by being somehow fitted into a genuine proof. But he derives the epistemic potency of computations by means that are radically different from the Fregean procedure just described. In the Fregean procedure, the basic tool is that of *semantical interpretation*. But on the Hilbertian model, the crucial device is not *interpretation* but *evaluation*. Under evaluation the constituent parts of a computation are *not* themselves endowed with any semantical significance and hence do not become the objects of noesis. Nonetheless, an evaluation *does* produce genuine judgements which form the basis for noesis with respect to the proposition expressed by the terminal formula of the evaluated computation; namely, the metamathematical judgements described earlier.

So, Frege and Hilbert agree in taking the epistemic potency of a computation to be ultimately based on a genuine proof that is somehow derived from the computation. However, the nature of the derivation, as well as that of the resultant genuine proof differ radically in the two cases. Yet both afford a solution to the first of the Fregean sub-problems: namely, the causal founding of noesis regarding the conclusion of a computation on other noetic events. In the Fregean cases, those other neotic events are noemata supposed to result from the *interpretation* of the constituent elements of the computation. In the Hilbertian case, they are noemata that are taken to be the result of metamathematical *evaluation* of the computation as a whole.

We must now say how it is that Hilbert's Metamathematical Replacement Strategy provides for the negotiation of the second of the two obstacles raised by Frege's Problem; namely, that of explaining how noesis formed via the Hilbertian mechanism might aspire to the *higher* forms of noesis. On the Fregean model this is accomplished by dint of the fact that the constituent premises and inferences of the genuine proof that results from the *semantical interpretation* of a computation are supposed to provide an *evidensory basis* for the proposition expressed by its conclusion. On the Hilbertian model, the constituent premises and inferences of the genuine metamathematical proof that arises from the *metamathematical evaluation* of the computation can, by reasoning parallel to that of the Fregean case, be seen as providing an evidensory basis for its (i.e., the metamathematical proof's) conclusion. And since the conclusion of that proof is just the proposition expressed by the terminal formula of the computation, it follows that the metamathematical proof provided by the Hilbertian model is capable of founding the higher forms of noesis regarding the conclusion of a computation. Hence we have a solution to the second of Frege's two sub-problems.[18]

In light of its nearly total silence on Frege's Problem, it is somewhat odd that the literature on Hilbert's Program should have given so much attention to a related problem. The problem that I have in mind, which we now describe but briefly, was first stated by Poincaré (cf. Poincaré [1908], pp. 169–171) and later, though independently, by Brouwer (cf. Brouwer [1912], p. 71). Because it is most commonly associated with Poincaré in the literature, and because he seems to have been the first to have actually formulated it, we shall call it "Poincaré's Problem".

In essence, Poincaré charged that Hilbert's Metamathematical Replacement Strategy cannot yield a satisfactory solution to

Frege's Problem (more particularly, to the second sub-problem) for more than a negligibly small portion of classical mathematical practice. This is so, on Poincaré's view, because classical mathematics is rife with applications of the principle of mathematical induction; and this holds no less of the ideal than of the contentual portion of classical practice. Therefore, given that Hilbert's metamathematical proof of the real-soundness of the ideal methods must itself invoke the principle of induction, the Hilbertian shall find himself in the position of using an application of induction to justify applications of induction. And this, according to Poincaré, begs the question of the reliability of those very methods whose reliability is at issue.

The Hilbertian response to Poincaré, which we can only sketch in this chapter, is as simple and elegant as it is powerful.[19] It consists in drawing a distinction between the use of induction in metamathematical evaluations of ideal proofs and its use in those ideal proofs themselves. The key idea is *not* that induction *per se* is different in mathematics than in metamathematics. Rather, it is that the one type of use (metamathematical) is contentual while the other (ideal mathematical) is computational or ideal, and that the logic of the former is importantly different from that of the latter. It is this difference between the logic of contentual metamathematics and that of ideal mathematics that allows an inductive metamathematical proof of the reliability of an inductive ideal proof to be non-circular. Such, at any rate, is Hilbert's view of how to handle Poincaré's Problem. We shall examine both the problem and the response more closely in the next chapter.

Let me now summarize the discussion of the last few pages. We have been considering the question of how the ideal method is supposed to serve as an epistemic aid. The view that has emerged is one according to which the ideal method is prized because of its potential for increasing the *extent* of our epistemic holdings. If use of ideal derivations can actually be used to obtain noesis, and if it produces it more efficiently than the exclusive use of contentual methods would, then we can expect the use of ideal methods to widen the scope of our epistemic holdings.

This view, as we have just seen, was challenged by both Frege and Poincaré each of whom argued, though in different ways, that the use of ideal proofs could not be expected to produce noesis at all. Frege's argument is apparently based both on the general principle to the effect that only noetic events can be expected to cause noetic events and on the more specific claim that the construction of an ideal derivation (or computation, in general) is not itself composed of noetic events. From these two premises we are supposed to conclude that an ideal proof of a real formula cannot be expected to issue in noesis with respect to the proposition expressed by that formula. Poincaré, on the other hand, anticipated the core of the Hilbertian response to Frege's Problem; namely, that while construction of an ideal derivation of a real formula will not *by itself* produce noesis in the proposition expressed by that formula, the contentual metamathematical *evaluation of* that ideal derivation will. But he failed to appreciate the difference between the contentual metamathematical deployment of induction and its ideal deployment. Hence, he was led to conclude that even the metamathematical evaluation of ideal derivations could not lead to any higher form of noesis in a very wide range of cases since the circularity of such evaluations would prevent this.

We have sketched the Hilbertian response to these objections. But even if one were to grant that response, and grant further that the use of ideal proofs makes such noetic acquisition more efficient than it would be if we were to use only contentual mathematical methods, it would only follow that the use of ideal derivations could be expected to widen the scope of our epistemic acquisitions. It would *not* follow that we ought to use the ideal method. For it might be that this increase in the quantity of our epistemic holdings should come about only at the cost of a decrease in their *quality*. And if such a qualitative loss were great enough, it would override the quantitative gains afforded by the ideal method, making its use, on balance, an epistemic liability. So the Hilbertian instrumentalist must concern himself not only with the relative efficiency of the ideal method, but also with the quality of its epistemic results.

We shall refer to this problem of quality-control as "the Dilution Problem". And since what we are here calling the "quality" of a given epistemic acquisition is mainly a function of the *strength* of the evidence for it,[20] the Dilution Problem, as we shall present it, shall be primarily concerned with the comparative *strength* of contentual mathematical proof and its proposed metamathematical replacement. If, on the whole, the proposed metamathematical replacements yield weaker evidence for the propositions proven than do the contentual mathematical proofs which they are to replace, then we say that the proposed replacement produces epistemic dilution. The Hilbertian instrumentalist must show that, on the whole, the noetic strength of contentual mathematical proofs is matched by that of their proposed metamathematical replacements.[21]

It is the Dilution Problem that moves the Hilbertian instrumentalist to embrace finitism. For the contentual mathematical proofs which are to be replaced under Hilbert's proposal are proofs of finitary character. Hence, at least on Hilbert's estimation, they are proofs that are constructed from evidence of unusual strength. As such, they are proofs that we cannot expect to replace without dilution unless the proofs with which we replace them are of equally exceptional strength. And while this does not strictly imply that our metamathematical replacements (and hence the proof of the soundness of the ideal methods on which they are based) need be finitary, it does make that an especially attractive strategy for dealing with the Dilution Problem.

So, it is to insure that his metamathematical replacements will match the special strength of the mathematical proofs which they are to replace that leads the Hilbertian instrumentalist to search for a *finitary* proof of the reliability of the ideal methods. And although a more detailed version of this argument will have to be worked out in Chapter II, the rudiments of the argument should be clear even now: (1) finitary evidence is evidence of a special strength, (2) the Hilbertian instrumentalist proposes replacing finitary mathematical proofs with metamathematical proofs, (3) if this replacement is not to result in epistemic dilution, the meta-mathematical replacements must be as strong as the finitary proofs that they are to replace, (4) the special strength of finitary evidence suggests that the surest (and, perhaps, the only) way of doing this is to have the metamathematical replacements be finitary themselves, and thus (5) it is only reasonable for the Hilbertian instrumentalist to seek a finitary proof of the reliability of our ideal methods on which to base his metamathematical replacements.

But this way of viewing matters should lead the Hilbertian instrumentalist to place certain restrictions on his own Replace-ment Strategy. For surely some finitary *mathematical* proofs will be so short and simple as to render them both more efficient *and* more secure than their proposed metamathematical replacements. In such cases, metamathematical replacement would not only fail to provide any *quantitative* noetic gain (because of loss of effi-ciency), it would also insure a qualitative noetic loss (i.e., dilution). Therefore, such replacement ought to be whole-heartedly opposed by the Hilbertian instrumentalist.

Hilbert registered such opposition in a much-neglected distinc-tion between the so-called problematic and unproblematic real propositions and proofs. Originally, the distinction was charac-terized so as to separate those real propositions and proofs that syntactically abide by the rules of classical logic (= the unprob-lematic reals) from those that do not (= the problematic reals).[22]

In mathematics, we found, first, finitary propositions that contain only numer-als, like

$$3 > 2, \qquad 2 + 3 = 3 + 2, \qquad 2 = 3, \quad \text{and} \quad 1 \neq 1,$$

which according to our finitist conception are immediately intuitive and directly intelligible. These are capable of being negated, and the result will be true or false; one can manipulate them at will, without any qualms, in all the ways that Aristotelian logic allows. The law of contradiction holds; that is, it is impossible for any one of these propositions and its negation to be simultaneously true. The principle of "excluded middle" holds; that is, of the two, a proposition and its negation, one is true. To say that a proposition is false is equivalent to saying that its negation is true. Besides these elementary propositions, which are of an entirely unproblematic character, we encountered finitary propositions of problematic character, for example, those that were not decomposable [into partial propositions]. (Hilbert [1925], pp. 380–1)[23]

But since the motive behind the distinction seems to be that of saying where the ideal methods can be gainfully applied and where they cannot, the spirit of the distinction is really somewhat more encompassing than the remark just quoted might suggest. According to this broader conception, it is intended to demarcate two classes of real propositions. The first class, the so-called unproblematic reals, possess real proofs that are so evident and efficient as to render them unsupplantable by ideal methods. The second class, the problematic reals, have only real proofs that are either so long or logically so complex that they lack the evidentness and efficiency which makes a real proof immune to displacement by ideal methods. Hence, the problematic reals should be seen as including not only real propositions whose contentual proofs formally deviate from the principles of classical logic, but also those which may abide perfectly by the laws of classical logic but whose only real proofs are of unwieldy length or complexity.

Thus, we should think of the problematic/unproblematic distinction as delimiting the instrumentalism of the Hilbertian; i.e., as showing which items of contentual mathematics may be gainfully replaced by ideal methods and which may not. And, clearly, the Hilbertian instrumentalist does not advocate the replacement of every piece of contentual mathematics. It is only for such contentual mathematical proofs as have greater length/complexity than their metamathematical counterparts that there is any instrumentalist motive for replacement. And since the proof of reliability which is part of every metamathematical replacement is clearly a lengthier and more complex piece of contentual reasoning than some contentual proofs found in mathematics itself, it follows that there is no basis in epistemic utility upon which to found an *unrestricted* mathematical instrumentalism.

There is, however, another reason why Hilbert advocates only a restricted form of mathematical instrumentalism. And, ultimately, it has to do with his belief in a fundamental parity between the methods of contentual mathematics and the methods of his proposed metamathematics. On Hilbert's view contentual arithmetic consists of thought-experiments with some sort of idealized numerals.

... the objects of number theory are for me ... the signs themselves, whose form may be generally and reliably identified by us independently of the place, time, and the special conditions of the production of the signs ..." (Hilbert [1922], p. 163)[24, 25]

The objects of elementary number theory are thus taken to be signs, but signs considered in abstraction from their actual location, time of inscription, color, size, chemical composition, etc.; i.e., signs treated from the point of view of their "form". Thus, the signs dealt with in the thought-experiments of number theory need not be actual signs. For the conditions of their actuality are abstracted away from in the thought-experiments dealing with them.

But, our immediate purpose is not so much to give an accurate description of the nature of finitary reasoning as it is to disclose the second basis for Hilbert's restriction of his instrumentalism. And his views regarding the nature of elementary number-theoretic reasoning are of concern to us only because they supply some background which serves to motivate this second agrument. The really crucial element of the argument is not the attribution of this or that particular character to number-theoretic reasoning, but rather the essential identification of its character with that of proof-theoretic or metamathematical reasoning. This Equivalency Thesis is asserted by Hilbert repeatedly.

... it is possible to obtain in a purely intuitive and finitary way, just like the truths of number theory, those insights that guarantee the reliability of the mathematical apparatus. (Hilbert [1925], p. 377)

To prove consistency we ... need only show that ... $0 \neq 0$ is not a provable formula. And this is a task that fundamentally lies within the province of intuition just as much as does in contentual number theory the task say, of proving the irrationality of '$\sqrt{2}$', that is, of proving that it is impossible to find two numerals x and y satisfying the relation $x^2 = 2y^2$, a problem in which it must be shown that it is impossible to exhibit two numerals having a certain property. Correspondingly, the point for us is to show that it is impossible to exhibit a proof of a certain kind. But a formalized proof, like a numeral, is a concrete and surveyable object. It can be communicated from beginning to end. That the end formula has the required structure, namely "$0 \neq 0$", is also a property of the proof that can be concretely ascertained. (Hilbert [1927], p. 471)

But if metamathematical reasoning is essentially equivalent to that which is found in elementary number theory, and if metamathematical reasoning must be treated as contentual in order to solve Frege's Problem, then elementary number theory must also be treated as contentual. In Hilbert's instrumentalism, then, what is to be counted as contentual mathematics is (at least partially) determined by what evidence one takes to be necessary for the metamathematical proof of reliability; and any reasoning of the same basic sort that is found in mathematics must then also be counted as contentual. Since Hilbert believed that one could prove the reliability of the ideal methods using only finitary reasoning, he also believed that the contentual portion of mathematics could be restricted to that which is finitary. But it seems clear that the argument just developed would have led him to revise his estimate of the extent of contentual mathematics (and, hence, his estimate of the limits of mathematical instrumentalism) had he become convinced that more than finitary evidence is required to proved the reliability of the ideal methods.

DIGRESSION. In light of the argument just given, there is a certain temptation to conclude that the finitism of the Hilbertian instrumentalist is to a considerable extent arbitrary and optional. But this temptation ought to be resisted, as our discussion of the Dilution Problem should remind us. As long as the Hilbertian program advocates the replacement of certain finitary proofs of contentual mathematics with metamathematical substitutes, dilution is a threat unless the metamathematical substitutes are themselves finitary. Thus it is that the threat of dilution provides the Hilbertian instrumentalist with a strong, clear-cut motive for metamathematical finitism. And this motive is not weakened by the fact that his identification of the boundary separating contentual from non-contentual mathematics is mediated, under the action of the Equivalency Thesis, by what he takes to be the fundamental characteristics of his metamathematical reasoning.[26] *End Digression.*

Hilbertian instrumentalism, as we have developed it here, bears comparing to the more recent brand of mathematical instrumentalism defended in Field [1980]. Indeed, it seems that Hilbertian instrumentalism stands to benefit from such a comparison; for it shows a sensitivity to Frege's Problem and the Dilution Problem and the workings of an Equivalency Thesis that is lacking in Field's account.

The ultimate purpose of Field's instrumentalism is to show that we can account for the applicability of mathematics to empirical concerns without having to count it as true, and hence (according to Field's reasoning) without having to grant an ontology of abstract objects with the attendant well-known epistemological difficulties.[27] And, like Hilbert before him, Field turns to metamathematics in framing his account of the applicability of mathematics to physical thought. But unlike Hilbert, Field offers a version of instrumentalism that is completely opposed to any Equivalency Thesis linking metamathematics and mathematics. He relies on a metamathematical proof of the reliability of the ideal apparatus to account for its applicability, but he is committed to shunning any suggestion that the evidence appealed to in the metamathematical proof of reliability is equivalent, in any important sense, to the evidence putatively required for a realistic treatment of mathematics. More specifically, he is committed to the view that the problem of evidence in metamathematics does not essentially go beyond that which is associated with a nominalist ontology. And it is this nominalistic end which his defense of instrumentalism is supposed to serve.

Field begins his defense of nominalism by saying that there is *only one* serious argument against nominalism, and that is the one suggested in various of Quine's writings, and given a sustained presentation by Putnam.[28] According to this argument, one must use mathematics in order to do science and to carry out ordinary inferences about the physical world (where "using" mathematics consists in employing mathematical theorems as apparent premises in inferences).

However, Field contends, one can account for the inferential utility of mathematics without interpreting mathematical sentences as propositions, but rather simply as tools of inference whose specific utility is to "speed" things up.

> ... even someone who doesn't believe in mathematical entities is free to use mathematical existence-assertions in a limited context: he can use them freely in deducing nominalistically-stated consequences from nominalistically-stated premises. And he can do this not because he thinks those intervening premises are true, but because he knows that they preserve truth among nominalistically-stated claims. (Field, [1980], p. 14)

> ... invoking real numbers (plus a bit of set theory) allows us to make inferences among claims not mentioning real numbers much more quickly than we could make those inferences without invoking the reals. And the inferences we make in this way will be correct every time. *Prima facie*, this might seem to be good evidence that the theory of real numbers (plus some set theory) is true: after all, if it weren't true, invoking it in arguments in this way ought to sometimes lead from otherwise true premises to a false conclusion. But we've seen . . . that this *prima facie* plausible argument is deeply mistaken: the fact that the theory of real numbers (plus set theory) has this truth-preserving property is a fact that can be explained without assuming that it is *true*, but merely by assuming that it is *conservative*, which is a different matter entirely . . . (Field [1980], p. 29)

So, the use of mathematical derivation is epistemically valuable in that it helps us to obtain nominalistically stated conclusions more efficiently than we could without its use. But this will further Field's nominalist aspirations only if it can somehow be shown that the proof of conservativeness on which Field bases his claim that mathematics is reliable is itself nominalistically acceptable. For, apparently, the only way that Field can solve Frege's Problem is via a mechanism like Hilbert's, in which the metamathematical proof of reliability must be taken contentually. However, if the usual standards of ontological commitment are employed, one *is* committed to abstract entities since such entities are referred to and quantified over in both the premises and conclusion of the proofs of conservativeness entertained by Field. Hence, the only apparent means available to Field for coping with Frege's Problem are antithetical to the larger nominalistic aims which avowedly motivate his instrumentalism.[29]

Field's failure to recognize the anti-nominalistic implications of his instrumentalism thus seems to stem from one of two sources: either he fails to clearly understand the character and seriousness of Frege's Problem, or he fails to fully appreciate the force of an Equivalency Thesis to the effect that the abstractness of the ontology of metamathematics is epistemologically equivalent to the abstractness of the ontology of mathematics (or certain portions of it).[30] In either case the result is clear: in order to avail himself of the alleged ontological/epistemological benefits for which he prizes instrumentalism, Field must forego a resolution of Frege's Problem; and in order to obtain a solution to Frege's Problem, he must relinquish his nominalism. This being so, Field's nominalistic instrumentalism seems fatally flawed.[31]

3. HILBERT AND QUINE: FREGE'S PROBLEM REVISITED

Unlike Field, then, the Hilbertian does not take his instrumentalistic account of the applicability of ideal mathematics as pointing toward a nominalistic philosophy of mathematics. But neither does he take the applicability of ideal mathematics as implicative of a realist philosophy of ideal mathematics. And he is, on this account, thoroughly at odds with the most popular argument for realism these days; namely, Quine's.

Defenders of Quinean realism typically base their case on two theses:[32]

(I)   The Indispensability Thesis — ". . . beyond a minimal level, we do not know how to do natural science without mathematics. In logician's jargon: any reasonably sophisticated natural scientific theory could be formalized only as an extension of some part of mathematics."

(II)   Duhem's Thesis — ". . . beyond a minimal level, no scientific hypothesis is tested individually; only relatively large and heterogeneous bodies of hypotheses are tested against experiment and observation."

From these two theses we are supposed to conclude

... that such mathematics as is required by natural science for making true predictions ... is confirmed thereby; there are no grounds for confining such confirmation to the natural science and denying it to the mathematical science.

As stated, however, the argument is tendentious. For it treats the mathematical part of science as being on an epistemic and semantical par with the non-mathematical part. Both mathematical theorems and physical hypotheses are treated as genuine propositions whose epistemic role is taken to be determined by their evidentness as truths. But for what reason? Surely applied mathematics would be just as reliable a guide to empirical truth if it were merely empirically sound or empirically conservative as it would be if it were literally true.[33] So, in order to account for its utility in constructing successful theories, we need not ascribe *truth* but only *truthfulness* to mathematics. And we can do this without assigning to it any literal semantical status.

Viewed in this way, Quine's argument for realism appears to be based on the same bias as Frege's Problem; namely, the belief that the only way that mathematics can become epistemically potent is by being *literally interpreted*. If this is so, then Hilbert's instrumentalist response (viz., that mathematics can become epistemically potent by being metamathematically *evaluated* rather than interpreted) would appear to count just as heavily against Quine's realism as it does against Frege's.

But some latter-day Fregeans (including some Quineans) have sought to add a new twist to Frege's Problem. George Berry, for example, charges that the instrumentalist cannot *explain why* mathematics works. And he suggests that this puts him at a relative disadvantage with respect to the realist.[34] And in his recent (and very valuable) study of Hilbert's Program, Dag Prawitz urges essentially the same point.

A reasonable foundation of mathematics cannot treat the transfinite [ideal] part of mathematics as an instrument, a black box, that happens to give correct results; the weakness of such a position ... is obvious since *the foundational task must be to explain why the instrument works*, i.e., *to understand it*. Of course, we do attach some meaning to the transfinite notions, which guides our formulation of transfinite principles, and in case this meaning is not sufficiently clear, the task must be to explicate it or extract its mathematically fruitful ingredients. In short, *to make Hilbert's program at all credible, one must require that it yields an interpretation of . . . the ideal sentences.* (Prawitz [1981], p. 268; emphasis mine)

Finally, Jonathan Lear applies what Prawitz says about Hilbert's restricted instrumentalism to the case of Field's unrestricted instrumentalism as well.

Merely to say that an arbitrary theory *T* is a conservative extension of our theory of the physical world *P* will not explain the usefulness or applicability of *T*. It would be easy to formulate a consistent theory *T*, prove that *P + T* is a conservative extension of *P*, and show that *T* is of no use whatsoever, in deriving consequences about the physical world. It is precisely because mathematics is so richly applicable to the physical world that we are inclined to believe that it is not merely one more consistent theory that behaves conservatively with respect to science, but that it is true. (Lear [1982], pp. 188–9)

In this newly modified form, then, Frege's Problem asserts that instrumentalism cannot *explain why* mathematics is a useful device for deriving truths because only an *interpretation of* (as opposed to an *evaluation of*) mathematics can do that. Apparently the challenge is no longer just that of describing an epistemic mechanism for instrumentalism which (1) specifies a scheme of events of the appropriate sort (viz., noetic events) to *causally explain* how noetic events could result from the use of the instrument, and which (2) shows that the beliefs which causally underlie the noesis produced by the instrument can also be seen as *evidentially justifying* it.[35] In a nutshell, it is not enough to *prove that* the instrument is reliable (which is what Hilbert appears to do), we must *explain why* it is reliable.

   But exactly what is it that the neo-Fregean is demanding? That is, what is it that an *explanation of* reliability is supposed to give us that a proof of reliability based on evaluation rather than interpretation never could? The legitimacy of the neo-Fregean critique of Hilbertian instrumentalism depends critically upon getting a clear answer to this question.

Part of what makes it difficult to see what the neo-Fregean realist is demanding is the poverty of his own realism. His criticism of instrumentalism takes the form of a comparison with realism. And this comparison is supposed to serve as a critique of instrumentalism by making manifest the superiority of the realist alternative. But the apparent advantage of realism is illusory since the mere interpretation of a formalism accomplishes nothing that cannot also be accomplished by its evaluation. Allow me to explain.

Presumably one paradigm for explaining how a formalism can serve as a useful guide to truths concerning a given subject-matter is to interpret the formalism in such a way as would enable us to see it as comprised of truths of that subject-matter. If F is made up of truths of subject-matter S, some of the mystery regarding F's remarkable utility as a guide to truths concerning S is surely removed.[36]

It is perhaps tempting to take this as the paradigm underlying the neo-Fregean realist's objection to instrumentalism. But closer scrutiny of his position shows that he falls far short of the ideal set forth in this paradigm. In essence, his failing is this: he can *interpret* mathematics (or ideal mathematics) and thus present it as a body of truths, but he has no apparent way to interpret it as a body of truths *of the subject-matter(s)* with respect to which it serves as a useful guide to truth.[37] The abstract objects of the realist's mathematical ontology form one subject-matter, the spatio-temporal objects of physics another.[38] And the realist has given us no more reason to believe that truth regarding the former should be a useful guide to truth concerning the latter than we have for believing that the biochemistry of bass livers should serve as a guide to predicting the stock market. The realist has not succeeded in making the subject-matter of mathematics one with the subject-matters with respect to which it serves as a useful guide to truth. Hence, he has failed to live up to the standards of his own paradigm and should judge himself just as harshly as he has judged the instrumentalist.

But there is another failure of the neo-Fregean point of view which is, to my way of thinking, even more disturbing. And that is its cavalier disregard of other paradigms for explaining the epistemic utility of a given formalism. Two paradigms in particular are worthy of further consideration.

The first of these, which we shall refer to as the naturalistic paradigm, takes as its leading theme the idea that a given method M of belief-acquisition might serve as a useful guide to truth in the hands of cognitive agents of a given type *not* because of any *a priori* or conceptual connection between truth and the features of propositions and/or theories favored by M, but rather because of certain *natural laws* relating special features of cognizers of that type with the order that they would know. In the hands of one type of cognizer seeking cognition of a given order, M might be an effective epistemic device, whereas in the hands of cognizers of a different type, M might be an epistemic disaster. And so, on this model, the success of M is to be explained not by citing some conceptual tie between the propositions favored by M and truth, but rather by citing certain lawlike connections between the users of M and the realm with respect to which they seek cognition. As such, it seems to be clearly antithetical to the paradigm of the neo-Fregean. For that paradigm sees the explanation of M's epistemic utility as proceeding from the location of *conceptual* links between truth and the features of a proposition on the basis of which M selects it for membership in the epistemic corpus.[39]

Now it is at least conceivable that some naturalistic account of the utility of mathematics (or ideal mathematics) treated as a formalism be worked out. And if this is so, then the neo-Fregean objection to instrumentalism (viz., that it *could* never afford an explanation of why the alleged instrument is useful) simply manifests an unjustified preference for one sort of explanation over another. But rather than develop this argument further, let us press on to elaborate a different model for Hilbertian instrumentalism.

This paradigm shall be referred to as the replicationist paradigm. At its heart lies an analysis of the notion of epistemic utility which sees the epistemic utility of an instrument as being composed of two distinct ingredients: namely, *efficiency* and *acuity*.

What we are here calling efficiency has to do with how rapidly or easily a device D generates conclusions about a given subject-matter. D's efficiency as a guide to truths concerning a subject-matter S will be determined by how many conclusions regarding S are generated by D per unit time and/or effort.[40] And we will say that device D_1 is more efficient than device D_2 as a guide to truths concerning S just in case D_1 generates more conclusions regarding S per unit time/effort than D_2 does.

On the other hand, what we are terming acuity is itself to be thought of as an amalgam of two elements: *perspicacity* and *reliability*. We shall say of an instrument D that it is a *perspicacious* guide to the truth regarding S just in case it succeeds in generating as outputs a significant body of the truths regarding S. And we shall consider an instrument D a *reliable* guide to the truths concerning S to the extent that the only conclusions regarding S that it generates are truths. Finally, an epistemic instrument is acute to the extent that it is perspicacious and reliable.

We have, then, a somewhat refined idea of what is required for the replicationist's explanation of the utility of a given epistemic device: we shall need, in the first place, an account of its efficiency and, in the second place, an account of its acuity (which, in turn, will require an account of its perspicacity and an account of its reliability). Of course there is no reason either to suppose or to require that a single feature of the device shall underlie *both* the explanation of its efficiency and the explanation of its acuity. No more is there a reason to suppose or require that a single property of the device will account for its acuity; perspicacity and reliability are, after all, quite independent of one another and it is easy to imagine devices that would have the one trait but not the other.[41]

The main idea behind the replicationist treatment of acuity is to present the instrument whose epistemic utility is to be accounted for as being *fashioned after* that subject-matter with respect to which it is a helpful guide. Furthermore, in order to account for its efficiency the instrument should clearly reflect those traits of human cognition which are responsible for its being an attractive epistemic device. The chief idea behind any instrumentalist proposal is, presumably, that there is some sort of dissonance between the way the mind works and the way the world works, so that thinking about the world only in terms of literal truths and literal inferences about it proves to be an epistemic disadvantage, while thinking about it nonliterally, in terms more accommodative of the way the mind works, proves to be an epistemic advantage.

There is nothing incoherent about the instrumentalist's picture. It is, in the end, simply that of a mind that is (1) relatively poorly equipped to cognize a given reality if all of its thinking about that reality must be literal, but (2) relatively well-equipped to cognize that reality if it is allowed to think in terms of some convenient fiction or to think not in a literal mode at all (either true or fictional) but in some sort of computational or calculary mode. Knowledge is thus conceived of as representing some sort of compromise between mind and reality.

Following his analysis of epistemic utility, the replicationist's strategy for solving the neo-Fregean problem is to break it down into three component tasks: namely, the explanation of the instrument's efficiency, its perspicacity, and its reliability. We have already examined the psychological account with which the Hilbertian replicationist proposes to explain the *efficiency* of ideal mathematics.[42] Furthermore, the critique of the neo-Fregean is not directed at the question of the efficiency of the ideal method, but rather its perspicacity and reliability.[43] Consequently, the remainder of our discussion shall be focused on the replicationist's treatment of these latter two questions.

The central idea of the replicationist's strategy for handling the acuity question, we have said, is to explain the epistemic utility of an instrument by showing it to be patterned after the subject-matter with respect to which it is taken to serve as a useful guide to truth. True to this idea, the Hilbertian replicationist accounts for the utility of ideal mathematics as a guide to the truths of real (i.e., contentual) mathematics by presenting the former as deliberately patterned after the latter. More precisely, ideal mathematics is taken to be modelled after real mathematics in two different senses: (A) real mathematics is embedded in ideal mathematics (i.e., *all* truths of real mathematics are theorems of ideal mathematics), and (B) real mathematics is contentually exhaustive of ideal mathematics (i.e., *only* truths of real mathematics are contentual theorems of ideal mathematics; in other words, ideal mathematics is real-sound). (A) is accomplished by taking ideal mathematics to be what you get when you "close" the axioms of real mathematics under classical rather than finitary logic. Since every finitary consequent of a set of real propositions is also a classical consequent of that set of propositions, ideal mathematics is guaranteed to contain every theorem of real mathematics. (B) is, of course, to be realized by means of a proof of the real-soundness of ideal mathematics or a proof of the conservativeness of ideal mathematics considered as an extension of real-mathematics.

By constructing ideal mathematics in accordance with (A), we provide for an account of the *perspicacity* of ideal mathematics considered as a guide to the truths of real mathematics. For in constructing ideal mathematics, we start with real mathematics as a base and *embed it in* the ideal instrument. Hence, the fact that the ideal instrument generates a significant body of real truths among its consequents is hardly to be wondered at. The perspicacity of an instrument is mysterious only if it is apparently constructed without reference to the subject-matter with respect to which it serves as a guide to truth. In keeping with the replicationist stratagem, the Hilbertian instrumentalist seeks to avoid this element of mystery by explicitly basing his ideal mathematics on real mathematics.

In like manner, deliberately constructing ideal mathematics in accordance with (B) provides for an explanation of its *reliability* as a guide to the truths of real mathematics. For if we make our ideal extension of real mathematics pass the test of real-soundness (which is, of course, what Hilbert explicitly required), then its reliability as a guide to real mathematics is readily explicable: since the standards of truth that hold sway in real mathematics can be used to verify every real consequent of ideal mathematics, every such consequent must be regarded as true when judged by those standards. The reliability of an instrument is mysterious only if it is constructed without taking measures to insure its fidelity to its subject-matter. The Hilbertian instrumentalist avoids such mystery by adhering to the replicationist ideal of fashioning the instrument after the contentual subject-matter with respect to which it (the instrument) serves as a guide. Only now, replicating the subject-matter in the instrument does not consist in embedding the subject-matter in the instrument (as it did in our account of the instrument's perspicacity), but rather in embedding the instrument in the subject-matter! So, we take measures to guarantee that the instrument doesn't generate claims about the subject-matter that are not independently sanctioned by its own contentual methods. And this is, of course, exactly what the Hilbertian's metamathematical proof of the reliability of ideal mathematics is designed to accomplish.[44, 45]

This, then, completes our description of the replicationist strategy and how the Hilbertian instrumentalist employs it to obtain an explanatory account of the epistemic potency (i.e., applicability) of the ideal method and hence to respond to the neo-Fregean objection of the Quinean realist. The strength of the replicationist strategy is that it analyzes the epistemic potency of an instrument into the component factors of efficiency, perspicacity and reliability with the idea that different features of the instrument will then be called upon to account for each of the three different factors. Following this idea, the Hilbertian instrumentalist seeks to (1) account for the efficiency of the ideal methods by citing a psychological predilection among humans for the classical logic in terms of which the ideal method is framed, (2) explain the perspicacity of the ideal method by showing how it is deliberately fashioned as an extension of real mathematics, and (3) explain the reliability of the ideal method by showing that it is deliberately constructed in such a way as to be a *conservative* extension of real mathematics.

It thus appears that the Hilbertian instrumentalist has at least set forth a coherent proposal that addresses every legitimate question regarding the epistemic potency of ideal mathematics as an instrument for acquiring knowledge of real mathematics. This being so, the neo-Fregean plaint of the Quinean realist appears to be without foundation when applied to the Hilbertian's instrumentalist treatment of ideal mathematics.

Of course, the replicationist strategy, as we have developed it here, only provides an epistemology for ideal mathematics relative to one for real mathematics. We have not provided an account of the applicability of ideal mathematics to the physical world. And, indeed, we cannot do so at this time. Still, we should like to offer an outline of how we think one might go about doing so.

We shall begin by adopting von Neumann's suggestion (cf. von Neumann [1931], p. 52) that even in its empirical applications, the utility of ideal mathematics is due to the fact that it serves as a short-cut for contentual mathematics. This, however, does not explain why ideal mathematics is empirically applicable, but rather reduces the problem of the empirical usefulness of idea mathematics to that of the empirical usefulness of real mathematics. So, in order to finally understand why ideal mathematics is empirically applicable, we must also have an explanation of why real mathematics is empirically applicable.

Our view is that, for purposes of explaining its empirical applicability, contentual arithmetic ought to be conceived of as an idealization of our experience regarding the arithmetical behavior of actual physical objects. In dealing with physical objects we are often concerned with questions of an arithmetical character; e.g., questions concerning what the count of the combination of two ensembles is, what the count of the remainder of a depleted ensemble is, etc. Of course, as we all know, the arithmetic behavior of physical objects is not invariant under all situations. If I take five rabbits from one pen and three from another and put them all into one pen, then, if I count the rabbits directly after putting them into the common pen, I will normally come up with eight. But if I wait six months before counting the rabbits in the combined cage, I may come up with fifteen. And this is but one of many illustrations (ranging from the mundane instability of rabbit-arithmetic to the mystifying instability of the arithmetic of loaves and fishes) that might be given.

What might be called *mathematical* arithmetic abstracts away from the features of the physical order that make its arithmetic behavior unstable. Hence, when it is applied to the physical order, it is with the assumption that the arithmetically destabilizing features of the physical order will not exert an influence under the circumstances of the intended application. Evidently, this is an assumption which holds in a wide variety of physical circumstances, otherwise, mathematical arithmetic would not be the widely applicable device that it is.

Of course, it is not to be supposed that every piece of mathematical arithmetic would find physical application even though, on the whole, it arises from our encounters with the physical. This is due to the fact that the abstractions with which mathematical arithmetic begins come to lead a life that is independent of their origins. Those propositions which can be verified by the same basic type of evidence or thought-experiment as those abstract propositions having a physical application must be admitted to be corpus of mathematical truths even if they themselves have *no* such application. Hence, even though we take the origins of mathematical arithmetic to lie in physical experience, we need not embrace a view of mathematics which limits it to what is physically applicable.

We have, then, a very rough sketch of how contentual mathematics, and hence ideal mathematics, is applicable to the empirical order. Much more would have to be said before this sketch could deserve to be called a philosophy of mathematics, but, hopefully, what we have said gives the Hilbertian a promising line along which to pursue such a goal.[46]

4. CONCLUDING SUMMARY

The basic strategies for explaining the epistemic applicability or potency of a given piece P of mathematics (e.g., a theory, proof, or computation) are, then, of two fundamentally different types: (1) the so-called realist strategies which see the applicability of P as being explicable only by the assignment of an interpretation or meaning to P, and (2) the so-called instrumentalist strategies which contend that P can become applicable through being *evaluated* even if it is not assigned a meaning. Hilbert's Program, we have argued, is the embodiment of a strategy of the second sort.

However, Hilbertian instrumentalism is marked by a number of features which serve to distinguish it from other forms of mathematical instrumentalism. Each of these distinguishing features arises, to one extent or another, from the acute sensitivity that Hilbertian instrumentalism shows to the two big problems facing the instrumentalist: namely, Frege's Problem (in either its original or its Quinean guise) and the Dilution Problem.

It is out of sensitivity to Frege's Problem that the Hilbertian both makes the ascent to a contentual metamathematics (from whose vantage ideal mathematics may be *evaluated*) and adopts the replicationist stance, according to which he responds to the neo-Fregean objections of the Quinean realist. And it is out of sensitivity to the Dilution Problem that he embraces finitism. For he sees in a finitary proof of the real-soundness of ideal mathematics the cleanest guarantee that substituting a metamathematical proof of a real mathematical proposition for a real mathematical proof of that proposition will not diminish the quality of its noetic basis.

Furthermore, one can see concern for the second of Frege's sub-problems as that which underlies the Hilbertian's interest in Poincaré's Problem. For Poincaré's Problem is an attempt to show that use of a metamathematical substitute for mathematical proof will tend to produce circular arguments; i.e., arguments incapable of providing a noetic basic capable of sustaining any of the *higher* forms of noesis.

Finally, I think that there is some connection between the Hilbertian's restriction of his instrumentalistic attitude to ideal mathematics and his concern for Frege's Problem. For we seem to be forced to count finitary reasoning as contentual in order to have any chance of getting a contentual metamathematics which, in turn, we require for a solution to Frege's Problem.

Thus it is that our approach to Hilbert's Program is dominated by our concern for Frege's Problem (in each of its various forms) and the Dilution Problem. In this chapter, the focus of our concern has been Frege's Problem. The remainder of the book shall, by and large, be devoted to a discussion of the Dilution Problem, which becomes acute because of Gödel's Second Theorem. The one exception to this is the first part of the next chapter where we add some much-needed detail to our discussion of Frege's Problem and the related problem of Poincaré.

NOTES

[1] In the same vein, though more recently, Kitcher writes,

. . . Hilbert divides grammatical sentences into those which have truth-values and those which do not, assigning sentences which express finitary propositions to the former class and ideal statements to the latter class. Ideal statements would be regarded as meaningless marks which are manipulated to facilitate our reasoning about finitary matters. The position is just that of a simple instrumentalist philosophy of science. . . (Kitcher [1976], p. 105; cf. footnote 7 at the bottom of the page whence this quote is taken for further elaboration.)

The reader may consult von Neumann [1931], Prawitz [1972] and Kreisel [1958] for additional statements along these same lines. And for a concise statement of some non-instrumentalistic interpretations of Hilbert, the reader should consult the excellent survey article, Prawitz [1981] (especially pp. 259–60; 265–70), and Gentzen [1936, 1938]. For a view similar in certain respects to the instrumentalistic interpretation and similar in other respects to the non-instrumentalistic interpretation, see Dummett [1973], p. 219.

[2] As we shall see shortly, von Neumann saw these two ends as intimately connected: the real propositions are, ultimately, what is applied to the empirical world. Thus, in generating attitudes towards real propositions, use of the ideal method is also supposed to provide a basis for the epistemic acquisition of empirical propositions.

[3] Of course, it might turn out that some propositions which are conjectured to belong to the extension of our intended notion of a true real proposition might not be provable by anything belonging to what we take to be the extension of our concept of a real proof. But were this the case, we would have reason to modify either our conjecture regarding the extension of our notion of a real truth or our conjecture regarding the extension of our notion of a real proof.

[4] For the time being, we assume that by reliability we mean real-soundness. We will say that an ideal proof of a real formula is sound if the real proposition expressed by that formula is true (according to the appropriate notion of truth, of course). A body of ideal proofs is real-sound if every real theorem of that body is true. Later on (in the remarks in Chapter IV on soundness and consistency) it will become clear that the issue of reliability is not quite so straightforward as it may seem.

[5] This is the passage referred to in note 2 wherein von Neumann connects the *empirical* reliability of our ideal methods with their reliability as guides to (real) mathematical truth.

Also, though von Neumann does not support this claim by mentioning examples, general support for it can be obtained from the fact that experience reveals that an ideal strategy like proof *reductio ad absurdum* can, at least in certain circumstances, give one a much shorter proof of a theorem than a direct proof of the same result. Gentzen, however, appears to challenge von Neumann's claim concerning the practical advantages of ideal proof; cf., Gentzen [1936], p. 170. This is obviously an important dispute, but not one that we shall attempt to settle here since (1) it is clearly a matter for the "experts" (i.e., practicing mathematicians) to decide, and (2) our major concern is to determine what must be done to justify the use of ideal procedures *granted* that they are of some practical value. For more on this, see note nine.

[6] According to Hilbert, there is a powerful impetus among humans to build thought into "units" of some sort.

... unser Denken geht auf Einheit aus und sucht Einheit zu bilden. (Hilbert [1930], p. 381)

[7] To view the development of an algebra or calculus of thought as an advantage is, of course, to suppose that sometimes it is an epistemic advantage for thought to proceed in a computational rather than a contential mode. But even within the computational mode, Hilbert makes a distinction: a computation that proceeds according to classical forms of logical derivation is, on the whole, more supportive of human epistemic acquistion than one that adheres to finitistic or non-classical forms of logical derivation. It is this superior effectiveness of *classical computation* that is supposed to give ideal proof its special place as an epistemic device of computational character. Hilbert readily admits that the real or contential reasoning that is to be displaced by ideal reasoning *can* be formalized (and indeed, his epsilon-elimination strategy for proving the soundness of ideal reasoning would *require* this). But he goes on to claim that even if it were, it would not provide a computational device as congenial to human epistemic use as one abiding by classical forms of formal inference.

[8] Weyl suggests something like this "Darwinist" line when he says that Hilbert's trust in the human psychological propensities which he took to be embodied in the procedures of modern classical mathematics, and which he directly argued for by appealing to their "practical success", must ultimately be based on faith in

... the reasonableness of history, which brought these structures forth in a living process of intellectual development ... (Weyl [1927], p. 484)

In making this observation, Weyl was neither criticising Hilbert nor questioning his assertion of the practical success of mathematics. Rather, he regards it as constituting a challenge to what he describes as "the philosophical attitude of pure phenomenology"; an attitude which, apparently, he takes to lie at the epistemological root of intuitionism.

9 This prima facie evidence, as we have already indicated, includes such considerations as (1) the fact that devices like non-constructive indirect proofs are, in many clearly envisageable instances, shorter than their finitary counterparts would be, (2) the fact that the systems of constructive mathematics that actually have been constructed appear to be much less easy to use as applied mathematics than the parallel ideal (i.e., classical) systems, and (3) the fact that classical as opposed to non-classical reasoning has been so dominant in the actual historical development of our own highly useful, fecund system of mathematics.

Perhaps the best examples of how ideal methods can be used to abbreviate and simplify contentual inquiry occur in that branch of mathematics known as Analytic Number Theory where asymptotic methods are brought to bear on combinatorial questions.

10 He put the point forward, however, in condemnation not only of what he took to be a crude version of formalism that he believed Hilbert to be committed to, but also, apparently, as a criticism of instrumentalism in general.

11 For Frege, a judgement is the acknowledgement of the truth of a thought (cf. Frege [1903], p. 86; [1928], p. 539).

12 Frege's use of the term 'proposition' is at odds with our own. What he here calls a proposition (i.e., a group of audible or visible signs) is what we call a formula. We take a proposition to be what one gets when a formula is interpreted. Hence, our own use of 'proposition' corresponds to Frege's use of the term 'thought'.

13 My use of the term 'noesis' is purely for the sake of brevity. I use it as a generic term to stand for the taking of any intrinsically valuable epistemic attitude toward a proposition, and (in particular) I intend no comparison with the way that Husserl used the term. Also, it should be noted that nothing essential to my argument depends upon my counting belief as a bona fide form of noesis (i.e., as an *intrinsically* valuable propositional attitude). For someone who believes that belief is *not* intrinsically valuable, Frege's Problem will just amount to the problem of determining how the construction of ideal proofs can be efficacious in the production of what I have referred to as the *higher* forms of noesis.

14 The reader should remember that the ideal proofs of which we speak are all supposed to have real propositions (or better, formulas *expressing* real propositions) as conclusions.

15 A more detailed treatment of Hilbert's resolution of Frege's Problem is given in Chapter II. In this chapter we are only trying to present a general outline of Hilbert's Program and its response to Frege's Problem. However, we should caution the reader that the Hilbertian response to Frege's Problem that

we outline here does not square perfectly with some things that have been said about the nature of finitary evidence (viz., that it is "logic-free"). My own belief is that Hilbert's Program cannot be accommodated within a logic-free conception of finitary thought, and so I oppose such a conception. But we shall reserve our more detailed remarks on the nature of finitary thought for the next Chapter.

[16] For this reason, we will often refer to Hilbert's instrumentalism as the "Metamathematical Replacement Strategy" (or, more briefly, the "Replacement Strategy").

[17] Actually, there is the potential for some confusion here. For what *we* are calling premises are not *judgements* acknowledging the truth of thoughts, but rather the thoughts themselves, i.e., what we have been calling propositions. Similarly, what *we* are calling inferences are not *judgements* of the truth-preservingness of shifts from one proposition (or set of propositions) to another, but rather the shifts themselves. This said, our present description of Frege's position should be seen merely as an attempt to state his views in the terminology that has already been established in this essay.

[18] Or at least as good a solution as the Fregean model itself provides.

[19] However, the power of Hilbert's reply seems not to have been appreciated in the literature where, most commonly, Poincaré's objection is presented as doing grave damage to Hilbert's Program.

[20] Of course, in a given epistemology, epistemic quality might not be taken to be exclusively a function of strength. One might, for example rank the propositional attitude of *understanding* (i.e., knowledge or justified belief *why*) as a higher epistemic good than that of knowledge or justified belief *that*. And, at the same time, one might be forced to admit that two bodies of evidence for a proposition P can be of the same strength (i.e., confer the same degree of certainty on our belief that P) even though only one of them *explains* P, and hence yields knowledge or justified belief *why P* holds.

In addition it may be that the ranking of our epistemic attitudes is not sensitive to all differences regarding strength. Thus, for a given attitude A, it may be that two pieces of evidence for P both afford the taking of A with respect to P even though they do not confer exactly the same degree of certainty on P. I am indebted to Steve Wagner for bringing a point similar to this latter one to my attention.

[21] This does not require that *each* metamathematical replacement be a match for the mathematical proof it replaces. However, establishing this stronger condition may be the only way to establish the weaker one just stated. At any rate, it seems to be the strategy favored by Hilbert for solving the Dilution Problem, as we shall see.

[22] Of course *no* real propositions semantically abide by the laws of classical logic (interpreted classically). But the unproblematic reals do syntactically or formally obey them.

[23] Actually, Hilbert specifically deals with two kinds of problematic real propositions; namely, those which are not decomposable into finitary subpro-

positions (the "partial" propositions, as he refers to them), and those he calls *hypothetical judgements*. He illustrates both of these with specific examples (cf. Hilbert [1925], pp. 377–8 for an example of the first sort, and pp. 378–9 for an example of the second sort). But each example generalizes nicely to provide a whole class of problematic real propositions. In general, a problematic real proposition of the first variety is a proposition of the form

(a)   There is an *F* greater than *p* but less than $p + k$ (where *p* and *k* are non-negative integers).

The problematic character of a proposition of this form (i.e., its failure to be inferentially manipulable by all the classically applicable formal modes of inference) is clearly illustrated by the fact that, *classically*, one can infer from (a) that

(b)   There is an *F* greater than *p*,

though, *finitarily*, one cannot, since (b) is not a real proposition. So, propositions of the general form (a) are real propositions, but real propositions of a decidedly different sort than $3 > 2, 2 + 3 = 3 + 2$, etc., since those latter propositions completely abide by the laws of classical logic while propositions of form (a) do not.

A problematic proposition of the second variety is a real proposition of the form

(c)   For every numeral *n*, *Fn* (where '*Fx*' is a formula such that when it is completely instantiated with numerals, it yields a proposition of the wholly unproblematic sort like $3 > 2, 2 + 3 = 3 + 2$, etc.).

Its problematic nature is revealed by the fact that the denial of (c) is not a real proposition, and so we cannot assert the classical law of excluded middle with respect to (c). But we can always do this for any proposition of the unproblematic sort (e.g., $3 > 2, 2 = 3, 2 + 3 = 3 + 2$, etc.).

[24] Hilbert and Bernays also characterize finitary reasoning as consisting in "Gedanken experimenten" with concrete, surveyable objects in Hilbert and Bernays [1935], p. 32. And in Kreisel [1958–9], p. 147 one finds substantially the same description.

[25] I think that whatever justification there is for labelling Hilbert a formalist comes from this passage and others making basically the same point. But it is clear that Hilbert was not a *crude* formalist who took mathematics to be a meaningless affair of shuffling symbols according to arbitrarily chosen rules. Rather, it is as Resnik says: Hilbert viewed contentual mathematics as a meaningful and useful theory of certain symbolic objects, and ideal mathematics as an instrumentalistic extension of contentual mathematics (cf. Resnik [1980], p. 54). We shall have more to say about the question of mathematics' usefulness later on.

[26] Of course, the traditional line on Hilbert's Program also takes his finitism to be neither arbitrary nor optional. But its motivation of finitism is radically

different from that which is given in this essay. It seems not to take notice of the Dilution Problem as a standing problem for instrumentalism. Hence its motivation of finitism makes no mention of the desire to avoid dilution. Rather, it sees finitism as stemming from a search for some sort of epistemological bedrock that would provide one with an *inalienable* guarantee of the reliability of any region of mathematics whose reliability is finitarily proven.

There are two variations on this theme: an older one in which finitary evidence is treated as infallible, and a newer and more sophisticated one formulated in Tait [1981] (and discussed in greater detail in the next chapter) in which finitary evidence is taken to be fallible though *unavoidable* and thus *not refutable*.

As will be seen in the next chapter, our treatment of finitism places far fewer demands on the strength of finitary evidence (and also on its homogeneity) than does this traditional view of Hilbert's Program. We do not need to assume either that finitary evidence is infallible or unrevisable, or that it is unavoidable, or that it is all stronger than non-finitary evidence, or, finally, that it is even all of basically the same strength (i.e., epistemically homogeneous). Each of these conditions seems to us to be implausible in one way or another. The condition that we shall place on finitary evidence, though itself not provably correct, appears to have much greater plausibility than any of the conditions just listed. As such it seems to us to constitute an advance in attempting to develop the epistemology of Hilbertian instrumentalism.

[27] These difficulties are recounted in Benacerraf [1973], where they are linked dilemmatically to the problem of giving a semantically satisfying account of mathematical truth.

[28] Cf. Putnam [1971], pp. 38–43.

[29] Interestingly enough, Field admits the platonistic character of the proofs of conservativeness saying,

... these proofs (at least the first, and both if one is sufficiently strict about what counts as nominalist) are platonistic, and so some story has to be told about how the nominalist is justified in appealing to them ... (Field [1980], p. 110)

And the only suggestion he offers for developing a nominalistically satisfying account of metamathematics is the following empirical stratagem.

... we've seen that the nominalist has various initial quasi-inductive arguments which support the conclusion that it is safe to use mathematics in certain contexts; if he then *using mathematics in one of those contexts* can prove that it is safe to use mathematics in those contexts, this can raise the support of the initial conclusion quite substantially. (ibid.)

But it is not clear that this would help, since basing one's belief on empirical and empirically verified mathematical evidence would not change the ontological commitments of the conservation theorem thus established. Nor would it eliminate those of the empirically verified mathematical premises used to establish conservativeness in the argument envisaged by Field. So it seems that the nominalist can draw no comfort from the tactic just described.

Field is now attempting to develop a nominalistic account of metamathe-matics by analyzing the key notions of proof-theory modally. I would like to thank Professor Field for sharing his unpublished writings on this subject with me.

[30] This Equivalency Thesis does not appear to be as strong as that which Hilbert advocated since it does not epistemologically equate metamathematical evidence with a certain body of mathematical evidence in *all* important respects, but only with respect to the *abstractness* of the objects reasoned about.

Perhaps it should also be noted that in comparing mathematics and meta-mathematics in this way we have made some assumptions that Hilbert would probably not like; e.g., that the semantics of mathematics and metamathematics is to be conceived of referentially rather than procedurally. In this essay we do not, in general, want to commit ourselves to referential semantics, but since such an assumption seems to be acceptable to Field, it should not be taken as weakening our criticism of his position.

[31] As we have already noted in passing, Field seems also to be oblivious to the Dilution Problem. And while we're not altogether sure how serious a threat dilution is under his conception of how nominalistic arguments are to be replaced by metamathematical surrogates, the remark quoted earlier (cf. note 29) where he suggests that the grounds for the metamathematical replacements is inductive or quasi-inductive seems to us to at least raise the specter of dilution.

[32] The statement of these two theses and the ensuing conclusion is taken from Hart [1979], pp. 154—5.

[33] By calling mathematics empirically sound or conservative, we mean that any empirical proposition derived from true empirical propositions with its aid can be derived from empirical truths without its aid as well; hence, any empirical proposition derived with its aid is empirically true (i.e., demonstrable from empirical truths).

[34] Cf. Berry [1969], p. 255. Berry's name for instrumentalism is "the abacus-theory" which he defines on p. 254. Berry's discussion suffers a bit from his failure to distinquish nominalism and instrumentalism as clearly as he might.

[35] These are the two sub-problems of Frege's Problem that were distinguished earlier in this chapter.

[36] But *only* some. After all, it is not to be expected that every set of truths regarding S will be just as useful as a guide to the truths of S generally as every other set. So, F's *special utility* as a guide to the truths of S would not be explained merely by showing F to be comprised of truths of S.

[37] This point was brought home to me during discussions of Lear's article with Tadashi Mino. It is a little surprising to me that Lear's antipathy toward instrumentalism persists despite his seeming awareness of this deficiency regarding the currently known varieties of realism. He sketches an antidote with respect to geometry, but gives us no idea of how to remove the deficiency in other areas of mathematics such as arithmetic and analysis. Nonetheless, he deserves credit for noticing what seems to have gone unnoticed by other realists.

[38] The parallel point in the case of Prawitz' objection to Hilbert's instrumentalistic treatment of the ideals is this: there is no known way of giving *the same sort of* interpretation (viz., finitary) to the ideals as Hilbert gave to the reals.

[39] Presumably, that is what is to be accomplished by *interpreting* the formalism that has been identified as a useful guide to the truth. By ascribing truth-conditions to the theorems of a formalism we pave the way for a demonstration that their truth-conditions provide for the satisfaction of the truth-conditions of those truths with respect to which they serve as a useful guide.

[40] The conclusions should, of course, also be suitably varied. But this is a detail that shall not detain us here.

[41] A device that generated all propositions regarding S would be perspicacious since it would generate all of the truths concerning S. But it would obviously not be reliable since it would also generate all of the falsehoods regarding S. Similarly, a device which generated only trivial truths of S would be reliable, but not perspicacious.

[42] Cf. the discussion with which notes (3)–(9) are associated.

[43] However, Lear's version of the neo-Fregean objection may be an exception to this claim. For Lear reasons that T's being a conservative extension of P does not explain T's usefulness as a guide to the truths of P on the grounds that not every conservative extension of P would be a useful guide to P. Of course T's being a *deliberately constructed* conservative extension of P *would* account for its perspicacity and reliability with respect to P. What it *wouldn't* account for is precisely its efficiency since not all conservative extensions of P can be expected to facilitate derivation of p-consequences at the same rate of effort and/or speed.

Lear apparently believes that the efficiency of T is ineluctably linked to its truth, since he goes on to suggest that it is only T's truth that can account for its usefulness as a guide to P. So his version of neo-Fregean realism seems to have a different basis than the others (which, presumably, would see T's truth not as necessary to account for its efficiency, but rather for its perspicacity and reliability). And it seems even less plausible than the others since efficiency seems to be the one component of epistemic utility which is *easiest* to account for without invoking truth.

[44] But here we must be careful, because the need to explain the reliability of ideal mathematics as a guide to real mathematics does *not* motivate the Hilbertian instrumentalist's demand for a *finitary* proof of reliability. Clearly there are measures falling short of finitary proof which would accommodate an explanation of why an instrument constructed in compliance with them is reliable. So, the demand for a *finitary* proof of real-soundness is to be seen as stemming not from a desire to explain why the ideal method is reliable but rather from a desire to deal successfully with the Dilution Problem (as was indicated earlier).

[45] At this point the realist might object that what is mysterious is not that an instrument successfully designed to be a conservative extension of P should be a reliable guide to P, but that it should be possible to construct a conservative extension of P that is not true when judged according to the (standard)

semantics of *P*. But such an objection reveals only an impoverished logical imagination or a poor understanding of what a conservative extension is.

[46] However, I would like to touch upon one particularly vexatious problem
which remains to be overcome. And that is the problem of explaining how one
sort of idealization leads to contentual mathematics while another sort of
idealization leads to non-contentual mathematics.

One might say that the abstraction which leads one to contentual arithmetic
does so because it remains within the bounds of what is determinable by a
constructive thought-experiment on objects treated combinatorially, but that
the abstraction or idealization which leads to ideal mathematics does not. But
this presumes the ultimacy of the constructive/non-constructive distinction as
determining the boundary between what has contentual status and what has
merely instrumental status. And it is not clear that the constructive/non-
constructive distinction is deep enough to serve in such a capacity.

Of course, the Hilbertian instrumentalist might just deny that there is any
sort of *qualitative* difference between the two types of idealization. That is, he
might admit ideal mathematics as a contentual theory and still maintain the
importance of getting a finitary proof of its reliability by appealing to the
Dilution Problem. Hence, although we have presented the Hilbertian instrumentalist as maintaining the non-contentuality of ideal mathematics, our treatment could be readily adapted to accommodate a view of the real/ideal
distinction which sees it as signifying only a quantitative distinction between two
types of contentual items.

CHAPTER II

A CLOSER LOOK AT THE PROBLEMS

1. INTRODUCTION

In the previous chapter we presented what we take to be the most serious challenges to the program of the Hilbertian instrumentalist. In this chapter we shall seek both to deepen our understanding of and develop (at least in outline) a response to these challenges.

The discussion of Frege's Problem will center on the issue of the Hilbertian instrumentalist's use of induction and of logical inference in general. And nowhere will our readiness to play the role of the revisionist be more in evidence than it is in this discussion. We shall argue that the Hilbertian must be willing to engage in some genuine inference in order to carry out his program. More particularly, a case will be made for saying that the Hilbertian must make room for using mathematical induction as a device for effecting genuine inferences. The fact that some of Hilbert's own (rather equivocal and unsettled) remarks on induction seem to be at odds with our proposal shall not deter us, since we see no other way to obtain a cogent version of Hilbertian instrumentalism. Still, even on our proposal, there is no evident need to depart from the oft-affirmed (e.g., in earlier times by Hilbert, and more recently by Tait) standard that all finitary evidence is formalizable in Primitive Recursive Arithmetic. We feel that whatever infidelity we show to the more traditional accounts of finitary evidence is surely mitigated by this fact.

The treatment of Poincaré's Problem is also novel, though in a rather different way. Its novelty consists in taking Hilbert's own response to Poincaré seriously. This is something that Hilbert's contemporary interpreters have been unwilling to do, and the result has been a series of elaborate, even Byzantine, attempts to distinguish different kinds of induction. Yet Hilbert's own solution is as simple and elegant as it is powerful. He makes no attempt to distinguish varieties of induction *per se*. Rather, he simply emphasizes a point that should have been clear all along; namely, that finitary logic is a sub-system of classical logic. And from this basic observation one can fashion a compelling response to Poincaré's objection.

The discussion of the Dilution Problem returns us to a consideration of the character of finitary evidence. However, our concern this time is with somewhat more basic epistemological issues regarding its evidensory stature. Our claim shall be that previous epistemologies for finitary evidence all seek to hold it to much higher evidensory standards than are needed. In order to motivate the Hilbertian's commitment to finding a finitary proof of the soundness of the ideal methods of classical mathematics, one need commit himself neither to the infallibility of finitary evidence (as is advocated in Kitcher [1976]) nor to its "Cartesian" indubitability (as is proposed in Tait [1981]). Rather, one need only maintain its optimality under certain conditions. What emerges is, we believe, a much more tolerable evidensory burden for finitary evidence, and one which clearly ties the Hilbertian instrumentalist's commitment to finding a finitary proof of soundness of the ideal methods to a very legitimate concern; namely, that of gaining a satisfactory solution to the Dilution Problem.

2. THE STATUS OF INDUCTION

We noted in the last chapter that Hilbert sometimes spoke as if general formulae do not express literal propositions, but rather serve only as propositional schemata; i.e. devices that come to express a genuine proposition only when specific numerals are substituted for the variables. Hilbert's own rather bewildering description is given in the following passage where he purports to show why generalizations must be regarded as problematic.

... we come upon a transfinite proposition when we negate a universal assertion, that is, one that extends to arbitrary numerals. So, for example, the proposition that, if x is a numeral, we must always have

$$x + 1 = 1 + x$$

is from the finitist point of view incapable of being negated. This will become clear for us if we reflect upon the fact that [from this point of view] the proposition cannot be interpreted as a combination, formed by means of "and", of infinitely many numerical equations, but only as a hypothetical judgment that comes to assert something when a numeral is given. (Hilbert [1925], p. 378)

Generalizations are problematic because, when negated, they don't yield a contentual proposition, but only an ideal (= transfinite) one. That is clear enough. But the reasoning behind this is what is perplexing. For it suggests that a generalization is itself not a genuine proposition. If a proposition is an assertion, then Hilbert's suggestion that a generalization comes to assert something only when numerals are substituted for variables amounts to saying that it is only the instances of a generalization, and not the generalization itself, that have propositional status. But if this is so we can only wonder at Hilbert's earlier reference to generalizations as universal assertions (i.e., assertions that extend to arbitrary numerals),[1] and his evident desire to strike a contrast between a generalization and its negation where negating a generalization is described as an operation that carries one *from* the realm of the finite *to* the realm of the transfinite or ideal. And our perplexity increases when we pause to consider other remarks by Hilbert where he speaks of generalizations as *contentual communications*.

When communicating, we also use letters, such as x, y, z, for numerals. Accordingly, $y > x$ is the communication that the numeral y extends beyond the numeral x. And likewise, from the present point of view, we would regard $x + y = y + x$ merely as the communication of the fact that the numeral $x + y$ is the same as $y + x$. Here, too the contentual correctness of this communication can be proved by contentual inference . . . (Hilbert [1925], p. 377)

There seem, then, to be two different views of generalizations that one might attribute to Hilbert. One is the view the generalizations are not themselves genuine propositions but only schemata of genuine propositions. And the other is the view that generalizations are genuine propositions which, like their instances, may serve to communicate some contentual assertion.

Associated with each of these different views concerning the nature of generalizations is a corresponding view regarding the nature of induction. On one view, induction is not a method of proof but rather a method of proof-schematization. And on the other view, induction is a method of genuine, contentual proof and not merely a technique of proof-schematization.

According to the former conception, one obtains a genuine proof of a proposition of the form '$F(\mathbf{n})$' from an inductive proof-schema of the proposition-schema '$F(x)$' by the successive "instantiation" of the proof-schema up to the term '$\mathbf{n}$' (which we take to be the nth element in some ordering of our terms). In the first step of this so-called "instantiation" procedure, we prove that $F(\mathbf{1})$ is true. In the second step, we prove '$F(\mathbf{1}) \rightarrow F(\mathbf{2})$'. In the third step, we infer that $F(\mathbf{2})$. We continue in this manner until we have a proof that "looks" like this:

> Step 1: Prove $F(\mathbf{1})$
> Step 2: Prove $F(\mathbf{1}) \rightarrow F(\mathbf{2})$
> Step 3: Conclude (from Steps 1 and 2) that $F(\mathbf{2})$
> . . .
> Step $2n-3$: $F(\mathbf{n-1})$
> Step $2(n-1)$: Prove that $F(\mathbf{n-1}) \rightarrow F(\mathbf{n})$.
> Step $2n-1$: Conclude (from Steps $2n-3$ and $2(n-1)$) that $F(\mathbf{n})$.

In this proof of $F(\mathbf{n})$ we have $n-1$ premises of the form '$F(\alpha)$', and $n-1$ premises of the form '$F(\alpha) \rightarrow F(\alpha')$' (where '$\alpha'$' is the term coming after 'α' in our ordering of terms), and $n-1$ applications of what looks to be *modus ponens*. Such a view of induction is suggested in Hilbert and Bernays [1934], pp. 298—9.

On the other conception of induction, one obtains a proof of '$F(\mathbf{n})$' not by inferring it from $n-1$ premises of the form '$F(\alpha)$' and $n-1$ premises of the form '$F(\alpha) \rightarrow F(\alpha')$' via $n-1$ applications of *modus ponens*, but rather by inferring it from a pair of propositions to the following effect: (1) if, starting with the first term in our ordering of terms, we substitute successive terms of our ordering for 'x' in '$F(x)$', then, no matter how far up our ordering we might go, a true proposition will always result,[2] (2) '$F(\mathbf{n})$' can be obtained by such a sequence of substitutions. Here, clause (1) is taken to be a *contentual* generalization established through the use of induction, and clause (2) is seen as a *contentual* recognition that '$F(\mathbf{n})$' falls under that generalization. Together, they amount to a contentual proof of '$F(\mathbf{n})$'. Here, as is pointed out in Tait [1981], knowledge of (1) depends upon having a grasp of the *concept* of a finite sequence and of how finite sequences are built up from the null sequence. And possession of such a generic concept of number appears to be necessary if one is to have induction as a means of proof rather than merely a proof-schematization.

But the problems concerning the status of induction go even deeper than what we have indicated thus far. And these additional problems all stem, to one extent or another, from the idea that finitary thought is supposed to deal exclusively with "concrete" objects. We shall now attempt to ascertain what this commitment to the concrete amounts to. Following that, we will apply what we have learned to our current question regarding the status of induction in finitary thinking.

Even the early descriptions of finitism focused on two attributes of finitary evidence: namely, its commitment to the concrete (as opposed to the abstract), and its allegiance to constructive rather than non-constructive modes of reasoning. Hilbert and Bernays described finitary reasoning as a form of mental experimentation with concretely conceived objects, where the experiments conducted consist in envisioning what happens to a concrete object when one applies certain constructive operations to it.[3] Viewed on a certain level, it would seem that every variety of constructive reasoning could be described as consisting in thought experiments wherein one performs various sorts of operations on objects of a given type. Consequently, it would appear that what distinguishes one variety of constructive thought from another are (i) different choices of objects on which the operations are to be performed, and (ii) different choices of operations with which to transform one's objects. It is apparently from some such perspective that Gödel derived what has come to be accepted as the standard way of distinguishing finitistic and intuitionistic reasoning.

... we have to distinguish between two strands in the finitary approach. Firstly there is the constructivist strand, whose content is that we may not count a thing as a mathematical object unless we can exhibit it or actually manufacture it by a construction. Secondly, there is the demand that the objects one makes assertions about, with which one carries out the constructions and which one gets by the constructions, should themselves be 'inspectable'. In the last resort this means that the objects must be spatiotemporal configurations, built up from elements whose nature is irrelevant apart from whether they are equal or distinct. (By contrast, the objects in intuitionist logic are meaningful assertions and proofs). (Gödel [1958], pp. 134–5)

This description of the difference between finitary and non-finitary forms of constructivist thinking is amplified by the following remark in which Gödel sets forth the distinction between 'inspectable' and 'abstract' constructions.

... we must count as abstract (not inspectable) those concepts which are essentially of the second or higher order. By this we mean those concepts which do not comprise properties or relations of concrete objects (as for instance combinations of symbols), but which are concerned with thought-constructions (proofs, meaningful propositions, etc.). The consistency proofs will use insights into their thought-constructions, insights which are derived not from the combinatorial (space-time) properties of the combinations of symbols which represent the proofs, but only from the *sense* of the symbols. (Gödel [1958], p. 133)[4]

So, the constructions of finitary thought register the effects of applying syntactical operations to spatio-temporal objects, while the constructions of the less elementarty varieties of constructivism (such as intuitionism) represent the application of semantical and epistemic operations to either concrete syntactical objects or to abstract semantical and epistemic objects.[5]

[N. B. Gödel's description of finitism might lead one to think that the mental experiments of the finitist are direct at sign-tokens (i.e., actual physical inscriptions whose identity essentially depends upon their location, color, size, chemical composition, time of inscription, and so on). But, as was already indicated in Chapter I, this is not the case. There we saw that Hilbert expressly stated (cf. Hilbert [1922], p. 163) that it is the "forms" of sign-tokens that are to be the focus of the finitist's attention. And, for Hilbert, the "form" of a sign-token is independent of its time, place, chemical composition, etc.[6] Thus, the objects of finitary thought are sign-tokens, but sign-tokens treated as having only a "form" or "shape." The scope of a given finitary operation is, therefore, never just one or another particular token, but rather the whole class of tokens sharing a given form.[7]]

We have, then, what is for our purpose, a clear enough understanding of the sorts of "objects" that can be subject of finitary constructions or operations. But we must still say something about the character of the operations that, according to finitism, may be performed on an object treated from the point of view of its shape. Gödel tells us only this much: the object that we get as "output" by performing a finitary operation on a concrete object as "input" is itself to be a concrete object. Furthermore, the process leading from the input to the output object is to be effective or constructive in character.

There are, however, potentially important differences separating constructive processes with respect to what might be called their "surveyability".[8] In particular, there are those effective procedures consisting of the iteration of an elementary operation where we can tell before executing the process *how many* applications of the constituent operations will be involved in going from the "input" object to the "output" object. On the other hand, there are pairs of objects that can be related by an effective process or operation where we cannot tell before completing the process how many iterations of the constituent operations will be required in going from the "input" object to the "output" object. (This is, of course, nothing other than of familiar distinction drawn between the so-called *primitive* recursive procedures, on the one hand, and the *general* recursive procedures, on the other. Primitive recursive operations require only bounded applications of the μ-operator whereas general recursive operations require unbounded applications of it.)

Since we have somewhat better knowledge of the particular iterative procedure linking "input" with "output" in cases of the former sort than in cases of the latter sort, there seems to be a fairly clear sense in which procedures of the former sort may be said to be better "surveyable" or "visualizable" than procedures of the latter sort. At least this is so if we count it as proper to say that one has a procedure for producing the "output" from the "input" not only when one has actually executed such a procedure, but also when one only has *a means of* executing it.[9]

This general intuitive description of the objects and constructions of finitary thought is said by Kreisel to sustain the following more detailed conception of it.

Finitist mathematics does not use the general notion of a constructive proof at all, in fact it might be said to avoid logical inferences (which involve an impredicative concept of proof) because it is restricted to purely combinatorial operations. In particular, the logical connectives have a purely combinatorial character since they are applied only to decidable formulae ... Universal quantifiers are not used at all except insofar as they can be replaced by free variables, e.g., not in the premise of an implication. Existential quantifiers are used as shorthand for a (constructive) function or functional ... Iterated implications are not used at all. (Kreisel [1958-9], p. 148)

In this passage, Kreisel is concerned with two related issues. The first concerns how we are to understand the role of logic or inference in finitary thought. And the second conerns how the role of logic in finitary thinking constrains the syntactical description of finitary thought. We shall, for the time being, focus on the former concern.

Kreisel's claim is that finitary thinking is essentially free of logical inference. Evidently, part of what lies behind this claim is a conception of logical inference which sees it as consisting in some sort of cognitive operation wherein one abstract entity, the premise (treated now as a genuine proposition), is transformed into another, the conclusion. So conceived, logical inference cannot be included in finitary thinking because the latter is restricted to those mental acts in which one concrete object is transformed into another. The mental act of transforming one concretum to another may be recognized by the finitist as proof of the effects of the transformation, but it is not itself a concrete object. Hence, it cannot enter into finitary thought as the object of a transformation.

> In (combinatorial) mathematical *practice* the act of recognizing that two expressions are identical is accepted as part of the data without further analysis ... Thus the only abstract objects which have a place in this theory are proofs, but they are not in turn the subject of combinatorial reasoning. Proofs, considered as mental acts, are clearly not finite configurations of concrete objects ... (Kreisel and Krivine [1967], p. 198)

According to the conception of finitary reasoning being described here, not even the "chain-of-conditionals" rendition of induction that we set forth earlier can be viewed as finitary. For, basic to our previous description of that mechanism is the use of genuine logical *inference* to carry us from a proof of '$F(\mathbf{n})$' and a proof of 'if $F(\mathbf{n})$, then $F(\mathbf{n}')$' to the conclusion '$F(\mathbf{n}')$', (i.e., the proof of '$F(\mathbf{n})$' and the proof of 'if $F(\mathbf{n})$, then $F(\mathbf{n}')$', when combined, yield a proof of '$F(\mathbf{n}')$'). And if finitary thought is to be logic-free in the way that Kreisel and Gödel seem to suggest, then the finitist could not execute the various steps of *modus ponens* called for in the chain-of-conditionals rendition of induction as we presented it. *Modus ponens*, after all, requires one to operate on *proofs* (i.e., to *combine* the proofs of '$F(\mathbf{n})$' and 'if $F(\mathbf{n})$, then $F(\mathbf{n}')$') to obtain another *proof* (viz., a proof of '$F(\mathbf{n}')$').[10]

Now one might opt for a logic-free interpretation of the chain-of-conditionals version of induction.[11] But there is little to be gained from doing so, because the basic problem regarding induction (namely, that of developing the epistemology of proof-schemata) is not addressed by such a tactic. And when that problem is addressed, we shall see revealed a deep-seated need for genuine logical inference on the part of the Hilbertian instrumentalist. Thus, in the end, we are lead to the conclusion that any account of finitary reasoning which sees it as logic-free is not adequate to the needs of the Hilbertian instrumentalist.

In attempting to develop an epistemology for the notion of a proof-schema we are immediately faced with a very basic choice: either we treat a proof-schema as constituting a proof itself (i.e., we take a proof-schema for a proposition-schema '$F(x)$' as itself being evidence for the truth of each instance of '$F(x)$'), or we treat of proof-schema not as a proof itself, but only as a set of directions for constructing a proof (i.e., we take a proof-schema for '$F(x)$' as *not* giving evidence for the truth of each instance of '$F(x)$', but only as a set of directions for finding such evidence in the case of each instance). We shall argue that only the first alternative is acceptable for the Hilbertian instrumentalist and that it entails a commitment to the use of at least some geniune logical inference.

There are two reasons why the Hilbertian instrumentalist cannot maintain that a proof-schema does not itself serve as evidence for the instances of the proposition-schema which forms its conclusion-schema. One of these derives from aims peculiar to the Hilbertian's program, but the other should convict both Hilbertian and non-Hilbertian alike. We shall begin with the peculiarly Hilbertian reason.

Consider a proof-schema of soundness for a given formal system T of ideal proofs. And suppose, as seems entirely reasonable, that this proof-schema proceeds via a use of induction. Its conclusion-schema is a proposition-schema, which we shall abbreviate as '$C(x)$', to the effect that if x is a proof-in-T of a real formula R, then R is provable in F (where F is taken to be a formalization of body of finitary thought). The position that we are currently considering is committed to saying that this proof-schema of soundness for T cannot itself be taken as evidence for any instance '$C(\mathbf{n})$' of '$C(x)$'. The only way, on this view, to get evidence for '$C(\mathbf{n})$' is to "instantiate" the proof-schema for '$C(x)$'. But what we end up with when we do this is not a proof of R, but rather a set of directions for transforming a given ideal proof (in T) of R into a real proof (in F) of R. And by the very reasoning that has been used to deny the status of proof to the proof-schema for '$C(x)$', it must now be maintained that even combining the instantiation of the proof-schema of '$C(x)$' with a proof that n is an ideal proof in T of R does not yet yield a proof of R. Rather, what it gives one is, at best, a set of instructions for constructing a proof of R. And that, according to the present position, is not at all the same as having a proof of R.[12]

On this view, then, one would a get of proof of R only by actually transforming a given proof in T of R into a proof in F of R.[13] But if this is so, then Hilbertian instrumentalism loses all point. For under such circumstances, one can only get a genuine proof of a real proposition R by actually constructing an object-level proof for R (which is what is represented by a proof in F of R). But, as was argued in the previous chapter, the whole idea behind defending the ideal methods of mathematical proof is to thereby avoid the necessity of constructing the cumbersome and laborious object-level proofs of the truths of real mathematics. The ascent to metamathematics is not supposed to lead us to construct object-level proofs of the truths of real mathematics, but rather to enable us to *avoid* doing so by providing a genuine, albeit metamathematical, substitute for such proofs.[14]

This, then, is the first reason why the Hilbertian instrumentalist cannot tolerate a view of proof-schemata which denies it the epistemic status of a proof. The second reason is more direct. In essence, it charges that it is simply incoherent to deny the status of proof to proof-schemata. The argument for this claim takes the form of a simple reductio.

Suppose that a finitary proof-schema P for '$F(x)$' does *not* serve as evidence attesting to the truth of every instance of '$F(x)$'. Then there is an instance '$F(\mathbf{n})$' of '$F(x)$' such that P gives us no reason to believe that '$F(\mathbf{n})$' is true. But, by definition, a proof-schema for '$F(x)$' gives directions for constructing a proof of '$F(\mathbf{n})$', and if this is so, then P must surely give one a reason to believe that '$F(\mathbf{n})$' is finitarily provable. However, if P gives one a reason to believe that '$F(\mathbf{n})$' is finitarily provable it follows from the constructive character of finitary truth that it also gives one a reason to believe that '$F(\mathbf{n})$' is true. But this contradicts our original assumption. We conclude, therefore, that a finitary proof-schema for '$F(x)$' is epistemically tantamount to a proof that every instance of '$F(x)$' is finitarily true. Furthermore, when we try to explain how a proof-schema for '$F(x)$' comes to serve as a proof of '$F(\mathbf{n})$', we are led to posit an inferential mechanism. Let us now consider why this is so.[15]

The basic epistemic feature of a proof-schema is that it provides evidence for the instances of its conclusion not by actually leading one to produce constructions of them, but rather by convincing one that such constructions *can* (modulo certain idealizations) be produced. But how does one become convinced that there is a construction for '$F(\mathbf{n})$' when he has not actually produced it, and has only a schematic description of how to do so? I believe that the key to answering this question lies in recognizing that not all of the specific properties of a putative construction for '$F(\mathbf{n})$' are germane to determining whether or not it is a genuine construction for '$F(\mathbf{n})$'. Some putative constructions are known to be genuine by their being identified only as a construction of a certain *kind*. Were this not the case, then a proof-schema for '$F(\mathbf{n})$' could never serve as evidence for '$F(\mathbf{n})$' at all. For a proof-schema attains its epistemic force by convincing one that every construction of a particular *kind* — namely, the *kind* set forth in the schematic description which constitutes the proof-schema — is genuine. Upon establishing that a construction of this kind can be given for '$F(\mathbf{n})$', one then *infers* that '$F(\mathbf{n})$' is true. Thus, the general form of the inferential mechanism by which proof-schemata achieve their epistemic force is this:

(1)   Every construction of kind K is genuine.
(2)   It is possible to produce a construction of kind K for '$F(\mathbf{n})$'.

_____

(3)   '$F(\mathbf{n})$' is true.

(Here, knowledge of (1) comes via the recognition of something's being a proof-schema.)

This, then, completes our argument for the claim that the Hilbertian finitist must countenance at least some uses of genuine logical inference. In brief, this is because he (as well as every other reasonable being) must count proof-schemata as providing evidence for their conclusions, and seemingly the only way to do this is via the inferential mechanism just described.

But where does all of this leave us with respect to the questions that prompted this discursus in the first place; namely, the Hilbertian's attitude toward mathematical induction and generalization, and his consequent response to Frege's Problem? Our analysis of a proof-schema implies that it is to be taken as providing evidence for a generalization; namely, that all instances of its conclusion-schema are true. Hence, we take a proof-schema to function epistemically as a genuine proof for a genuine *general* proposition. As a consequence, we are committed to taking mathematical induction as a means of genuine proof and to seeing the associated generalizations as expressions of genuine propositions. Finally, this view of the nature of induction and generalization enables us to give the simple and straightforward solution to Frege's Problem that was sketched earlier in this section; namely, that the cumbersome O-proofs of a real proposition R possessing an efficient ideal proof in the system T may be replaced by a genuine, albeit metamathematical, proof to the following effect (i.e., and M-proof).[16]

(1) T is real-sound (by an inductive metamathematical proof).

(2) R is provable in T (by construction of an ideal proof in T of R).

---

(3) R is true.

Of course interpreting finitism thusly, sets us at odds with the traditional view of the finitist which sees him as dealing exclusively with "concrete" objects in a logic-free manner. For those who cling to this more traditional view, the argument of this section may be taken as a critique of Hilbertian finitism since, in my view, the Hilbertian finitist cannot give an adequate response to Frege's Problem so long as he maintains such a view of the nature of finitary thinking.

But if solving Frege's Problem requires a revision of the traditional views regarding the nature of finitary evidence, such a revision might also destroy any basis that the Hilbertian instrumentalist might have for preferring a *finitary* proof to a non-finitary proof of the soundness of the ideal methods. For it might eliminate all significant differences between finitary evidence and other sorts of evidence having a potential for such use. We don't believe that this is so, and we conclude this section by suggesting an argument to that effect.

Finitary evidence has been traditionally identified, at least extensionally, with what can be formalized in Primitive Recursive Arithmetic. Other attempts to revive Hilbert's Program have suggested that we allow the defense of the classical methods to make use of constructive evidence going well beyond such bounds. We shall not, and nothing that we have said thus far commits us to doing so.[17]

Still, our reasons (though they are not really "ours" but rather those that are sketched in Tait [1981]) for taking PRA to form a "bound" on finitary reasoning diverge clearly from those of the traditional account. On the traditional account, PRA is identified with finitary reasoning because it (PRA) is taken as being the mathematics of the "concrete". On Tait's account, finitary reasoning contains a certain amount of abstract thinking (namely, that which involves the *concept* of a finite interation of a concrete operation), but such abstraction does not go essentially beyond what is to be found in PRA (according to any coherent account of the role of induction and generalization in PRA). What makes finitary evidence special, on Tait's view, is that it represents the *minimum* commitment to abstract thinking that must be made. If one renounces the uses of abstraction that occur in finitary thinking, then one must also foreswear the development of any non-trivial body of mathematical thought, according to Tait's account. This is taken to be so on the grounds that every non-trivial body of mathematical thought *presupposes* such uses of abstraction.

With some modifications, we accept this account. And later on in this chapter we shall augment it with certain other conjectures to obtain what we take to be the epistemological view of finitary evidence required to motivate the Hilbertian's quest for a finitary vindication of the ideal methods. But for now, our aim is the more restricted one of indicating to the reader how the inclusion of some abstract forms of thought within finitary reasoning does not, by itself, force one to abandon either the traditional claim that all finitary reasoning is arithmetizable in PRA or the traditional claim that finitary evidence enjoys an exalted epistemic status. We believe that the argument from Tait just sketched carries us some distance toward that aim. Consequently, we do not believe that the solution to Frege's Problem that we are proposing in this essay forces us to renounce all the more important traditional claims regarding the character of finitary evidence, even though it does oblige us to make room for some abstract reasoning within finitary thought.

3. POINCARÉ'S PROBLEM

But there are worries concerning the Hilbertian's use of induction that are independent of the Fregean worries dealt with in the previous section. In fact, the best-known worry concerning induction — namely, that which was first formulated by Poincaré — is of exactly this sort.

As Poincaré saw it, Hilbert's metamathematical use of induction was ultimately bound to result in too much circular reasoning (cf. Poincaré [1908], pp. 169—71[18]). Ideal mathematics, on Poincaré's view, is replete with applications of induction (in one guise or another), so that that part of ideal mathematics which doesn't use induction is relatively trivial and unimportant. Thus, in order to instrumentally justify any appreciable portion of ideal mathematics, the Hilbertian will have to prove the real-soundness of ideal uses of induction. But since such a soundness proof will itself appeal to induction, the Hilbertian will find himself in the position of using induction to establish the soundness of induction. And such reasoning, in Poincare's judgement, is viciously circular.

This argument of Poincaré's is seriously fallacious. But in order to see its weaknesses clearly we need a more careful and exact formulation of the argument itself. Hence, it is to this task that we now turn.

What is standardly taken to make an argument circular is that every reason for doubting the conclusion is an equally strong reason for doubting one or more of the premises or inferences which lead up to the conclusion. (In speaking of doubt regarding a premise or a conclusion, we mean doubt regarding its truth. And in speaking of doubt regarding an inference, we mean doubt concerning its truth-preservingness.) Hence, circularity is defined only for contentual arguments. Ideal proofs, not being composed of genuinely contentual premises and inferences cannot, on the standard definition, be circular. Nor can a contentual use of induction beg the question of the truth of an ideal use of induction, since the latter is not contentual. What could happen, at least conceivably, is that a contentual use of induction beg the question of the *real-soundness* (as opposed to the truth) of an ideal use of induction. Hence, it is this sort of claim that we shall interpret Poincaré as making.

> P(1)   Any reason for doubting the real-soundness of I's use of induction is an equally strong reason for doubting the truth (truth-preservingness) of any contentual meta-mathematical induction that might be used to prove I's real-soundness.[19]

But even if P(1) is granted, Poincaré's argument still will not go through unless one also grants that the only basis for doubting the real-soundness of I as a whole is doubt regarding the real-soundness of its use of induction. Hence, Poincaré requires an additional premise to the following effect.

> P(2)   Any reason for doubting the real-soundness of I is an equally strong reason for doubting the real-soundness of its use of induction.

From P(1) and P(2), and some sort of transitivity principle for reasons, we may derive Poincaré's Claim.

> (PC)   Any reason for doubting the real-soundness of I is an equally strong reason for doubting the truth (truth-

preservingness) of any metamathematical induction that might be used to prove I's real-soundness.

Our response to Poincaré's argument consists of an attack on both P(1) and P(2). And we shall begin with the case aganist P(2), which is the less central and less novel of our two counter-arguments.[20]

Typically, an ideal proof that contains an instance of the induction schema as a premise will also contain other premises that are independent of the induction schema. This being so, it is clearly impossible to reduce all worries regarding the real-sound-ness of an inductive ideal proof to worries concerning the real-soundness of its inductive premise. The real-soundness of an ideal proof using induction is, after all, still partly determined by what its non-inductive premises are. Hence, we can be no more certain of the real-soundness of an ideal proof that uses induction than we are of the real-soundness of its non-inductive elements. And as a result of this, it follows that the second premise of Poincaré's argument is not generally true.

The first premise of the argument is also objectionable. And this, in effect, is how Hilbert responded to Poincaré. In Hilbert's view, one who accepts this premise is guilty of something tanta-mount to equivocation. Such a premise is tempting only to one who fails to properly distinguish what we call induction in the case of a metamathematical proof of soundness from what we call induction in the case of an ideal proof.

... as my theory shows, two distinct methods that proceed recursively come into play when the foundations of arithmetic are established, namely, on the one hand, the intuitive construction of the integer as numeral (to which there also corresponds, in reverse, the decomposition of any given numeral, or the decomposition of any concretely given array constructed just as a numeral is), that is, contentual induction, and, on the other hand, formal induction proper, which is based on the induction axiom and through which alone the mathe-matical variable can begin to play its role in the formal system.

Poincaré arrives at his mistaken conviction by not distinguishing between these two methods of induction, which are of entirely different kinds. (Hilbert [1927], pp. 472–3)

Hilbert's point, then, is that the contentual induction which figures in our metamathematical induction is radically different from the formal or ideal induction that is used in ideal proofs.[21] And because of this difference, the former may be used to establish the real-soundness of the latter without begging the question. But since Hilbert does not go on to detail specifically which aspect or aspects of the difference between contentual and ideal induction it is that enables him to avert Poincaré's objection, we must try to do this ourselves.

The chief point, as we see it, is this: the logic of finitary reasoning is a sub-logic of that of classical reasoning. So, even if finitary reasoning calls upon induction in order to establish the consistency of classical reasoning that employs induction, this need not amount to circular reasoning. The set of propositions that results from closing a given proposition (or set of propositions) under the rules of finitary logic is a subset of that which results from closing it under those of classical logic. Hence, it is at least conceivable that a use of induction under classical manipulation should result in an inconsistency while that same induction, manipulated finitarily, should not. And in light of this, it is clear that reducing the consistency of a classical use of induction to that of a finitary use of induction need not be circular.[22]

In sum, then, our criticism of Poincaré's argument is that both of its premises are deficient. The first is unfounded and overlooks the very important differences separating finitary from classical logic. And the second is false since it ignores the fact that ideal uses of induction typically occur in proofs having additional ideal propositions as premises. Taken either together or singly, these criticisms should lead one to conclude that Poincare's argument is radically unsound and that it lacks the cogency necessary to pose a serious threat to Hilbertian instrumentalism.

4. THE DILUTION PROBLEM

According to the Hilbertian instrumentalist, there is an overall advantage in replacing real object-theoretic, mathematical proofs of problematic real propositions (or "O-proofs", as we are calling them) with meta-theoretic proofs of those propositions (or "M-proofs"). The Dilution Problem, as it was outlined in the previous chapter, is the problem of showing that the gains in epistemic expansion that are promised by this replacement are not neutralized or indeed overcome by a corresponding drop in the general quality of our epistemic holdings. The cleanest way of doing this would be to show that the M-proofs themselves are a qualitative match for the O-proofs which they are to replace. In this way, one would avoid the tricky business of having to compare a quantitative noetic gain with a qualitative noetic loss in order to assess the overall epistemic value of Metamathematical Replacement.

However, if one is to show that the replacement of O-proofs by M-proofs does not result in epistemic dilution, then one must show that the M-proofs themselves are finitary. Such, at any rate, is the position of the Hilbertian instrumentalist. This position is derived from the simple observation that the O-proofs are finitary and the claim that finitary evidence is evidence which possesses special strength. For if the O-proofs, being finitary, possess unusual strength, then the threat of dilution is especially acute when we attempt to replace them with M-proofs. Thus, it would appear that the best hope of dealing successfully with the Dilution Problem is to make sure that the M-proofs (and, hence, the proof of the real-soundness of the ideal methods on which the M-proofs are based) are themselves finitary. That way, we shall be replacing proofs of a type known for their special strength *with* proofs of that same type. And that would seem to minimize the chance of dilution.[23]

Now I do not wish to deny that, in part, the Hilbertian's commitment to finding a finitary defense of the ideal methods should be taken as reflecting certain foundational concerns. It is, after all, only reasonable that he should seek to build as strong a case as is possible for the reliability of the ideal method. And if a finitary soundness proof would contribute to this aim, then the Hilbertian should by all means try to obtain one.

However, we wish to emphasize that the foundational concern is not the only, nor even the strongest, motive for seeking a finitary proof of the reliability of the ideal methods. There is also the Hilbertian's unavoidable concern with the Dilution Problem. And, in our view, this is a point that needs to be emphasized both because of its absence from the literature on Hilbert's Program, and because of the salutary effect it should have on our views regarding the basis and depth of the Hilbertian's commitment to finding a finitary soundness proof.[24]

Of course, the reasonableness of the Hilbertian's concern with the Dilution Problem will be determined by the plausibility of his claim regarding the special status of finitary evidence. For it is that claim of special status which gives the Dilution Problem its edge. We should like, therefore, to determine what status must be attributed to finitary evidence in order to motivate the Hilbertian's finitism. Our finding is, I believe, somewhat surprising. For, as we shall soon see, the Hilbertian's commitment to a finitistic meta-mathematics can be derived from a much more plausible view of finitary evidence than is familiar to one through the literature on Hilbert's Program. To see that this is so, let us give ourselves briefly to a consideration of two recent studies of Hilbert's Program which seem to me to span the range of alternatives available in the literature. The studies I have in mind are those of Kitcher [1976] and Tait [1981].

According to both Kitcher and Tait, Hilbert is supposed to have held that finitary evidence is infallible, or capable of providing absolute certainty. In Kitcher's opinion, Hilbert held such a view because he wanted to show that the epistemological distinctiveness of mathematics lies in the unique certainty of its evidence (cf. Kitcher [1976], p. 99). For Tait, though he is not explicit on the point, Hilbert presumably maintained the infallibility of finitary evidence because he wanted to create a foundation for classical mathematics that would last forever. Furthermore, both Kitcher and Tait hold the view that Hilbert's deep epistemology for finitary evidence was conceived along basically Kantian lines. The reckoning of "shapes", which is the basic form of judgement in finitary reasoning, is taken to be conceived by Hilbert as a form of sensory or para-sensory intuition which is not reducible to anything more basic. This immediate intuition, in turn, is taken to be that which Hilbert sees as providing the special security of finitary evidence.

Now, both Kitcher and Tait agree that such a Kantian attempt to provide a deep epistemology for finitism is a failure. And their central contention is that one cannot provide an account of how generalizations (e.g., $x + 1 = 1 + x$) are known if all one has to go on is the representability in intuition of a number or a number-sign. As Tait puts it,

We discern finite sequences in our experience . . . We not only discern such sequences but we see them *as* sequences, i.e., as having the *form* of finite sequences. I shall call this form *Number.* . .

There are subforms of Number, for example, even number, square number, etc. The minimal such subforms are the numbers. O is the form of a null sequence, 1 of a one-element sequence, and so on. Thus the relation of Number to the numbers is not that of a concept to the objects falling under it but that of a form or structure to its least specific subforms.

We do not understand Number via the concept of number, i.e., of being a number. Rather, it is the other way around. We understand the numbers as the specific determinations of Number. For example we see

| | | | | | | | | | | | | | | | | | | | | | |

as a finite sequence. Then we count and determine its number. If it were otherwise, what would we be counting? How would we choose the units and their order? We are able to see the above inscription as a sequence in many ways. Also, by suitable reading, e.g., conventions, we could see it as a sequence of any number of units, each in many different ways. Again, we understand $n+2$ not via understanding each of the infinitely many instances, $0+2$, $1+2$ and so on. Rather, we understnad these via our understanding of what it means for one sequence to be a two-element extension of another. (Tait [1981], pp. 529–30)

Can the notion of an arbitrary two-element extension (or any other arbitrary k-element extension) be given in intuition? Tait thinks not. For in order to do so, we would need the notion of an *arbitrary* number, i.e., we would need to understand the idea of Number. And, as Tait observes,

The real difficulty . . . is that the essence of the idea of Number is iteration. However, and in whatever sense one can represent the operation of successor, to understand Number one must understand the idea of iterating this operation.

But to have this idea, itself not found in intuition, is to have the idea of number *independent of any sort of representation in intuition.* (Tait [1981], p. 539)

So, the evidence for generalizations cannot be just what is representable in intuition. Hence, the following argument (which both Kitcher and Tait attribute to Hilbert)[25] cannot be maintained since its second premise is false.

P(1)   Pure, Kantian intuition (and only pure Kantian intuition) is infallible.

P(2)   All finitary evidence is reducible to pure, Kantian intuition.

Therefore,

(C)   Finitary evidence is infallible.

There are, of course, other reasons for opposing an eipistemology of infallibility for finitary evidence than disagreement with P(2). But since these are well-known, I will not pause to reiterate them here. I turn instead to a more interesting proposal broached in Tait [1981].

Tait explicitly denies that a status of absolute certainty can be claimed for finitary evidence. Nevertheless, he maintains that it is optimal (as certain as any evidence can be) and, indeed, indubitable. On Tait's account, this special status of finitary evidence is to be seen as ultimately deriving from its unavoidability.

... the special role of finitism consists in the circumstance that it is a minimal kind of reasoning presupposed by all nontrivial mathematical reasoning about numbers. And for this reason it is *indubitable* in a Cartesian sense that there is no preferred or even equally preferable ground on which to stand and criticize it. (Tait [1981], p. 525)

Tait's argument, insofar as it is valid, echoes Hilbert's point regarding the unavoidability of finitary reasoning (cf. Hilbert [1927], p. 465, and note 25). Tait explicitly says only that finitary reasoning is presupposed by all nontrivial mathematical reasoning. But in order to reach his conclusion of Cartesian indubitability, he must either expand this claim to cover *all* nontrivial rational thought (and not just nontrivial *mathematical* thought), or he must maintain that every basis from which finitary reasoning might be critized would presuppose some nontrivial mathematical thought and hence some finitistic thought.

My guess is that Tait would prefer the latter. But the question of which auxiliary premise he would prefer is not the question which is of most concern to us here. For regardless of which of the two auxiliary premises is chosen, a serious problem remains. And this problem concerns the ambiguity of Tait's conclusion. When he says of finitary reasoning that *it* is indubitable, what does he mean? More specifically, does he mean to imply that if we found a finitary proof of the soundness of the ideal methods that *it* would be indubitable? I think that he does. But such an assertion is a highly problematic one for Tait to make, as I will now attempt to show.

Let us begin by distinguishing four different types of unavoidability which we will refer to, respectively, as weak global, strong global, weak local, and strong local unavoidability. We will say that finitary evidence is globally unavoidable in the weak sense if, whenever K is a nontrivial body of propositions, there is some element of K (or set of elements of K) that presupposes some piece F of finitary evidence. And finitary evidence will be said to be globally unavoidable in the strong sense when every nontrivial body of propositions K is such that for every piece of finitary evidence F, there is an element of K (or set of elements of K) that presupposes F. Finitary evidence is locally unavoidable in the weak sense if, given that K is nontrivial, for every element e of K, e presupposes some piece F of finitary evidence. And finally, finitary evidence will be said to be locally unavoidable in the strong sense if, for every nontrivial set of propositions K, there is some one piece F of finitary evidence such that every element of K presupposes F.[26]

Now Tait faces the following task. He must first use some thesis regarding the unavoidability of finitary evidence to establish a thesis asserting its Cartesian indubitability or optimal security. He must then employ this indubitability thesis to found a conclusion to the effect that a finitary proof of the soundness of the ideal methods would make *it* (i.e., the soundness of the ideal methods) indubitable. We do not believe that Tait can accomplish this task, and we shall now try to show why.

The basic problem is this: the unavoidability theses that are strong enough to enable Tait to reach his objective are implausible, and the plausible unavoidability theses are not strong enough to afford attainment of the objective. The weak global and weak local options yield the most *plausible* variants of an unavoidability thesis. Unfortunately, neither is strong enough to enable Tait to secure his objectives. On the weak global model of unavoidability, only *some* piece of finitary evidence need be regarded as unavoidable (viz., that which is presupposed by the weakest nontrivial body of propositions), and therefore only *some* piece need be regarded as indubitable in Tait's Cartesian sense. One gets no *general* thesis asserting the unavoidability of finitary evidence as a whole. Because of this, one cannot proceed further to conclude that the evidence required for a finitary proof of the soundness of the ideal elements is unavoidable and hence indubitable. There is simply no reason to suppose that the finitary evidence that is unavoidable on the model of weak global unavoidability includes the evidence required for a finitary soundness proof.

Much the same holds of the weak local model of unavoidability. From it we can infer no conclusion to the effect that all finitary evidence is unavoidable. Hence, we cannot use it to establish the indubitability of finitary evidence generally. Furthermore, we have no separate grounds for saying that the finitary evidence rendered unavoidable (and hence indubitable) by the weak local model, though not inclusive of all finitary evidence, nonetheless suffices for the proof of the soundness of the ideal methods. Consequently, Tait's objectives exceed the resources of weak local unavoidability.

Basically the same sort of assessment obtains even in the case of the much less plausible strong local model. For even if one were to succeed in establishing the strong local unavoidability, and hence the Cartesian indubitability, of a particular piece F of finitary evidence, one would still not be assured of the unavoidability and indubitability of the particular piece of finitary evidence that might be required to prove the soundness of the ideal methods. Hence, even a thesis of strong local unavoidability would not yield an indubitability thesis strong enough to provide Tait access to his ultimate objective.

The model of strong global unavoidability shows more promise, but in order to assess it, we must first try to get clearer on what Tait means by Cartesian indubitability. He glosses the notion by saying that piece of evidence P is Cartesian indubitable when "there can be no preferred or even equally preferable standpoint from which to launch a critique" of P (cf. Tait [1981], p. 546, and the passage from p. 525 quoted earlier). And, given this way of understanding Cartesian indubitability, Tait's ultimate objective would require showing that a finitary soundness proof for the ideal methods would be a piece of evidence such that there could be no preferred or even equally preferable piece of evidence with which it might conflict.

But now notice that one can establish such a claim from a premise of the strong global unavoidability of finitary evidence only if one is also willing to make a pretty strong claim asserting the epistemic uniformity of finitary evidence. According to such a thesis, every piece of finitary evidence must be taken to be just as strong or secure as every other piece. Or, at the very least, no piece of finitary evidence can be taken to be stronger or more secure than that which might be required for a finitary soundness proof. Without such a thesis, the possibility that the finitary sound-ness proof should be found to conflict with a more potent piece of finitary evidence would remain open. And in such a situation, the stronger or more potent piece of evidence would, in effect, constitute a vantage from which to criticize the finitary soundness proof. Hence, without a strong epistemic homogeneity thesis, finitary soundness proofs will fail to live up to Tait's standard for Cartesian indubitability even if the strong global unavoidability of finitary evidence is accepted. And this means that Tait must rely on such a thesis in order to secure his goals.

However, I find no basis for believing in such an epistemic similarity or homogeneity thesis. The so-called strict finitists, it seems to me, are clearly right to maintain that the security of finitary evidence is not invariant. Increases in length and complexity, to mention but two conspicuous factors, diminish the security of even finitary arguments.[27] And Hilbert himself apparently intended to mark such variations in security by means of the problematic/unproblematic distinction which, as we saw in Chapter I, he used to charcterize finitary proofs and propositions.

Bernays, also, has noted, in an ironic tone, that each step of a consistency proof carries one both closer to and further from a proof of consistency (cf. Bernays [1935], p. 210). It carries one closer to a *completed argument* for consistency (and hence a proof of consistency). But it carries one away from a *demonstration* of consistency because the security produced by an argument is inversely proportional to its "length" and "complexity".

Citing this observation of Bernays' with approval, Kreisel writes:

Hilbert sometimes speaks of the reliability (Sicherheit) of finitist reasoning. As Bernays has pointed out . . . , realistically speaking, almost the opposite is true, the chance of an oversight in long finitist arguments of metamathematics being particularly great. (Kreisel [1958], p. 161)

And elsewhere, he criticizes what he identifies as formalism by saying,

. . . formal operations are distinguished by the property that only a particular kind of understanding is needed to carry them out.

The error of the formalist *doctrine*, which restricts itself to formal operations, is to assume that this kind of understanding has a particularly high *degree* of reliability. (Kreisel [1970], pp. 18—19)

DIGRESSION. It would, however, be a mistake to take Kreisel as claiming that there is nothing epistemologically special about finitary evidence. For on another occasion he has emphasized just such a point; claiming that one thing that makes finitary reasoning important for foundational questions is its combinatorial character. This character, says Kreisel, makes finitary reasoning special. For, at least on occasion (viz., when it isn't too long or logically too complex), it is so elementary as to be on an epistemological par with our most elementary cognitions; e.g., our simplest sense preceptions (cf. Kreisel and Krivine [1967], p. 198; Kreisel [1965], p. 151).

What Kreisel does insist on is the *uneven quality* of finitary evidence. Finitary arguments, like all arguments , weaken as their length and/or complexity increases. Because of this, one argument is not automatically superior to another just because it is finitary. Nor is finitary evidence all of the same strength. This is Kreisel's point, and it seems to me to be correct. *End Digression.*

So, Tait's claim of Cartesian indubitability for finitary evidence generally, and for a finitary soundness proof in particular, appears to be mistaken. However, in rejecting both Tait's account of the epistemological character of finitary evidence as well as the previously discussed infallibilist account, we need not give up, as hopeless, the search for a cogent epistemology for finitary evidence. For there is a status ascribable to finitary evidence which does not call forth damning criticism. And, at the same time, it confers enough privilege upon the position of finitary evidence to provide motivation for the Hilbertian's pursuit of a finitary soundness proof.

The special status for finitary evidence that the Hilbertian requires can be provided without going anywhere near a claim either of infallibility or of Taitian indubitability. All that is needed is the following claim which, hereinafter, we refer to as the Principle of Weak Optimality (or the "PWO", for short).

> (PWO) Other strength-affecting factors such as length and complexity being equal, a piece of finitary evidence is stronger than a piece of non-finitary evidence.[28]

The PWO motivates the search for a finitary soundness proof in this way: unless we are given reason to suppose that the desiderata of the *ceteris paribus* clause of the PWO (e.g., simplicity plus brevity) significantly favor a non-finitary proof of soundness, we can expect the class of O-proofs that are replaceable without dilution by M-proofs to be greatest when a finitary proof of soundness serves as the basis of the M-proofs.[29] We need not say that it is inconceivable that a non-finitary proof of soundness might do the job as well. But, under the force of the PWO, if we don't have a reason to believe that finitary soundness proofs are markedly lengthier, more complex, etc., than their non-finitary counterparts, then we shall expect that the class of O-proofs replaceable (without dilution) by M-proofs will be at a maximum when the latter are based on a finitary soundness proof. And this, in our view, is incentive enough to motivate the quest for a finitary soundness proof.

Of course, to argue fully for the PWO, we would have to know far more about human cognition than we do today. At the very least, we would need some substantial body of statistical evidence suggesting that the reasoning which figures in the thought experiments of the finitist is more reliable than that which goes to make up other, more embracing, forms of constructivist reasoning of the same degree of length, complexity, etc. And, so far as I know, there simply has been no systematic attempt to assemble such a body of evidence. Consequently, we can offer no conclusive defense of the PWO at this time. Such credentials as it has reside in such ordinary facts as that (1) it does not offend against anything that we know or suspect to be true, and (2) it seems to enjoy fairly widespread popularity among expert mathematicians. But we wish to claim only two things for the PWO: namely, (a) it is substantial enough to provide us with a sharp and forceful way of motivating the Hilbertian instrumentalist's quest for a *finitary* proof of the reliability of the ideal methods, and (b) it offers a more compelling account of the ultimate character of finitary evidence than has heretofore been advanced. Taken together, these two claims provide the basis for a new understanding of the Hilbertian's commitment to finding a finitary soundness proof (namely, one which links it to concern over the Dilution Problem). And it does so on the strength of an epistemological claim for finitary evidence which, though not conclusively established, is nonetheless far more plausible than what we would be led to expect from an acquaintance with the previous literature on finitism.

Still, we have only shown why the Hilbertian should desire a finitary soundness proof. We have not shown that such a proof is possible. And, as is well-known, this is precisely where the most serious challenge to Hilbert's Program, namely the Gödelian challenge, makes its entry. Both because of the power and because of the influence of this challenge, we shall devote the remainder of this essay to its evaluation.

NOTES

[1] When Hilbert calls a generalization a universal assertion, he uses the phrase "allgemeine Behauptung". And when he speaks of a proposition (or a universal generalization) as asserting something, he uses the verb "behaupten". So, apparently, there is no equivocation here on the use of the word "assertion".

[2] Perhaps the hypothetical character of this clause (as reflected in its use of "might") provides us with a way of regarding Hilbert's so-called "hypothetical judgements" as genuine judgements rather than judgement-schemata while still capturing their hypothetical character.

[3] Cf. Hilbert and Bernays [1934], pp. 32–3 (second edition); Hilbert [1925], pp. 376–7; and Hilbert [1927], pp. 465, 471. According to Bernays (cf. Bernays [1967], p. 502) the need to distinguish finitism and intuitionism was not seen until the mid-thirties when, through the work of Gentzen and Gödel, intuitionistic number theory came to be seen as (classically) equivalent to classical number theory. If one believes that finitary number theory is (classically) weaker than its classical counterpart, then one must conclude from this that finitism and intuitionism are not equivalent.

[4] The remark about consistency proofs here accompanies an earlier claim by Gödel that Bernays has convincingly argued that Gödel's Second Theorem shows that in order to prove the consistency of the standard mathematical theories, we must use "abstract" concepts.

[5] According to Kreisel [1965] (cf. p. 119), Gödel was the first to characterize the difference between finitism and intuitionism in this way.

[6] Elsewhere, Hilbert seems to identify the "form" of a sign-token with its "shape". In describing finitary thought he says that

... what we consider is the concrete signs themselves, whose shape, according to the conception we have adopted, is immediately clear and recognizable. (Hilbert [1927], p. 465)

At least part of the function of a "form" or "shape" is to mark the syntactical category (for the logical as well as grammatical syntax) of a given sign-token.

[7] I think that this is what Charles Parsons is trying to say in his more careful description of the distinction between intuitionism and finitism.

. . . intuitionism is not the only possible constructivist development of mathematics. . . A weaker and more evident constructive mathematics can be constructed on the basis of a distinction between effective operation with *forms* of spatiotemporal objects and operation with general intensional notions, such as that of proof. Methods based on operations with *forms* of spatiotemporal objects would approximate to what the mathematician might call elementary combinatorial methods or to the "finitary method" which Hilbert envisaged for proofs of consistency. (Parsons [1967], p. 205; emphasis mine)

[8] See Kreisel [1965], pp. 168–71 for an interesting attempt to lend a little more precision to the notion of a "surveyable" process. Takeuti [1975], pp. 81–96 also gives a worthwhile and more detailed discussion of the "surveyability" of various inductive procedures. Finally, Webb [1980], Chapter III gives a deep discussion of this and related issues.

[9] Perhaps it is this difference between primitive and general recursivity that prompted the idea that primitive recursive arithmetic be used as the formal codification of finitary reasoning. Cf. Hilbert and Bernays [1934], chapter seven, and Tait [1981].

[10] I am, of course, describing a constructive version of *modus ponens*.

[11] On this conception, Primitive Recursive Arithmetic is to be formulated as a logic-free equation calculus in which the "logical" operators are purely syntactical operators. According to such a conception, the conditional operators in the conditional premises of a chain-of-conditionals would be taken as signifying a schema of substitutions whereby a system of equations whose last equation is '$F(\mathbf{n})$' can be tranformed into a system of equations whose last equation '$F(\mathbf{n}')$'. See chapter III of Goodstein [1957] for more details on how the logical operators may be represented arithmetically.

[12] If having a set of directions for constructing a proof *were* taken as tantamount to having a proof, then a proof-schema, which just is a set of directions for constructing a proof, would have to be treated as tantamount to being a proof. But this is exactly what the present position is supposed to deny.

[13] Even then it is not clear that one would get a proof of the sort that is capable of addressing Frege's Problem, since a proof in F of R is just a formal object and not anything that is endowed with evidensory status simply by virtue of the means by which it is produced from the ideal proof in T of R.

[14] Hereinafter we shall refer to the object-level proofs of propositions of real mathematics as O-proofs, and to their proposed metamathematical replacements as M-proofs.

[15] This will vindicate our earlier claim that the Hilbertian instrumentalist must allow at least some place in his contentual thought for genuine logical inference.

[16] Of course, the mere cumbersomeness of R's O-proofs will not necessarily warrant its replacement by an M-proof. For the M-proof of R is itself cumbersome to a degree since it involves the proof of T's soundness (which, on Hilbert's schema of classification, would make it a problematic real proof). Thus, an M-proof of R would be used to solve Frege's Problem only in those circumstances where it (the M-proof) is less cumbersome than every known O-proof of R.

[17] However, we conceive of PRA (= Primitive Recursive Arithmetic) as a genuine theory using (at least sometimes) logical operators to signify epistemically genuine uses of logic, and not as a logic-free equation calculus.

[18] Somewhat later, Brouwer made essentially the same objection (cf. Brouwer [1912], p. 71).

[19] Here, and throughout this discussion, 'I' is to be taken as standing for an ideal proof which contains a use of ideal induction.

[20] An objection similar to our argument against P(2) may be found in Steiner [1975], p. 141.

[21] However, as we will see shortly, the important difference between ideal and contentual induction is not that the one is formal (ideal) and the other informal (contentual), but rather that the logic of the one sort of reasoning (ideal) is apparently more permissive than that of the other.

[22] But some wag will say "Wasn't Hilbert's own strategy for proving the consistency of the ideal system T one of showing that if finitary reasoning is consistent, then the classical T is also? And if one were to succeed in such an attempt, wouldn't it show that any reason to doubt T's consistency is an equally good reason to doubt the consistency of finitary reasoning? And if this in turn were the case, then wouldn't Poincaré be vindicated?"

Such reasoning would make every valid argument "circular". *Before* a reduction of T's consistency to that of finitary reasoning, one might very well have reasons for doubting T's consistency that are not equally strong reasons for doubting the consistency of finitary reasoning. Indeed, Hilbert's attempt to "reduce" the question of T's consistency to that of the consistency of finitary reasoning was prompted by precisely such a consideration. Had he been successful, he would have at least partially "de-activated" those extra reasons for doubting T's consistency; and that was the whole point of attempting the reduction. But "de-activating" the reasons would not show that they never existed or that they were irrational.

[23] It would not, however, altogether eliminate it. For, as we shall argue shortly, finitary proofs are of varying strengths. So, even if the M-proofs are finitary, replacing O-proofs with them might still produce dilution (if the O-proofs thus replaced are, as a rule, stronger).

However, faced with such a situation, the Hilbertian instrumentalist might just refuse to replace those O-proofs whose replacement threatens dilution. So long as he could show that the Metamathematical Replacement Strategy promises significant gains in the expansion of our epistemic holdings *even when* such O-proofs are *not* replaced, his program would retain its motivation.

[24] We find no solid evidence that would indicate that Hilbert himself was concerned with the Dilution Problem. But, in our opinion, he should have been. And if he had been, he would have been led to new motives for his quest for a finitary soundness proof.

[25] Actually, I think that Hilbert comes closer to maintaining the optimality than the infallibility of finitary evidence. His major claim regarding the strength of finitary evidence seems to be that it is

... requisite for mathematics and, in general, for all scientific thinking, understanding and communication ... This is the very least that must be presupposed; no scientific thinker can dispense with it, and therefore everyone must maintain it, consciously or not. (Hilbert [1927], p. 465)

We shall soon see that Tait himself offers an account of finitary evidence similar in spirit to this one.

[26] What I have to say about unavoidability claims holds for a variety of different notions of presupposition ranging from a weak one which identifies presupposition with entailment to a more robust one according to which one proposition P_1 presupposes another proposition P_2 just in case P_1 cannot be rationally accepted unless P_2 is accepted. Also, I am assuming that what I refer to as "a piece of finitary evidence" can somehow be represented as a proposition.

[27] See Kreisel [1958–9], pp. 148–9, and Wang [1970], pp. 39–41 (where he describes strict finitism under the heading "anthropologism") for characterizations of strict finitism. Also, see Dummett [1959, 1975], and Wright [1982] for interesting and useful discussions of strict finitism.

[28] The strength-affecting features of the *ceteris paribus* clause are all supposed to be features which stand to affect the strength of finitary and non-finitary arguments to the same extent when present in the same degree.

[29] The reader should take note of the fact that we are tacitly assuming that certain non-finitary methods (e.g., intuitionistic methods) might still be counted as *contentual methods* (i.e., methods of genuine proof) by the Hilbertian. In general, it seems that any constructive method can be countenanced by the Hilbertian as contentual even though not all constructive evidence is on an epistemic par. What is supposed to separate finitary evidence from other forms of constructive evidence is its allegedly superior strength, and not its being the only conceivable *contentual* method of thought (though Hilbert believed that one didn't *need* to countenance any other form of contentual thought in order to found mathematics).

CHAPTER III

THE GÖDELIAN CHALLENGE

1. INTRODUCTION

The primary goal of this chapter is to give a careful statement of what, for want of a more descriptive title, we refer to as the *Standard Argument* against Hilbert's Program (or the SA, for short). This argument, as the title of the chapter suggests, is that which is derived from Gödel's Second Incompleteness Theorem (or G2, as we shall refer to it from here on out).

After stating the argument with what may strike the reader as unusual and, perhaps even unnecessary care, we locate three serious weaknesses in it. The location of these weaknesses is what justifies the unusual care exercised in stating the argument, since it is only through such a careful statement of the SA that these weaknesses become apparent.

We lack the space to be able to give, in this introductory section, a precise and understandable statement of any of these weaknesses. But this much we can say: they point up the need for a new type of generalization or strengthening of G2 if the SA is to prove capable of refuting Hilbertian instrumentalism. The strengthening called for by the first weakness (which we refer to as "the Stability Problem") is one capable of showing that the set of properties of a formula which make it a fit expression of consistency for an ideal system subsumes a set of properties guaranteed to make it unprovable in that system (given that the system is consistent). And the strengthening demanded by both the second and third weaknesses (which we refer to, respectively, as "the Convergence Problem" and "the Problem of Strict Instrumentalism") is basically this: show G2 to be obtainable in an anti-Hilbertian form not only for the usual sort of formal system T, but for every formal system capable of reproducing the *humanly useable* portion of T.

In this chapter we shall only state these problems and say enough about them to convince the reader that they constitute serious weaknesses in the SA. A more thorough examination of them will be given in Chapter IV (where the Stability Problem is discussed) and Chapter V (where the Convergence Problem and the Problem of Strict Instrumentalism are taken up). But even in just *stating* these problems clearly, we probe more deeply into the relationship between G2 and Hilbert's Program than the previous literature on the subject has done.

2. THE STANDARD ARGUMENT

What we will refer to here as the original version of the SA is that variant of it that is based upon the original version of G2, in which Gödel constructed *a particular* formula of T (call it $\mathrm{Con}_G(T)$) and showed that

(1)   if T is consistent, then $\mathrm{Con}_G(T)$ is not provable in T.

Judging on the basis of how $\mathrm{Con}_G(T)$ was constructed, Gödel also conjectured that

(2) $\mathrm{Con}_G(T)$ "expresses" (in some sense yet to be clarified) the consistency of T.

And since it is generally agreed (at least on the usual conceptions of finitary evidence) that

(3)   every finitary truth (in particular every truth of the finitary metamathematics of T) can be "expressed" as a theorem of T,[1]

it is therefore tempting to conclude that

(4)   if $\mathrm{Con}_G(T)$ is not provable in T, then $\mathrm{Con}_G(T)$ does not express a theorem of the finitary metamathematics of T.

This done, we are led to say (appealing to (2) and (4)) that

(5)   if $\mathrm{Con}_G(T)$ is not provable in T, then the consistency of T is not provable in the finitary metamathematics of T.[2]

And with (5) at our disposal we may call upon (1) to get

(6)   if T is consistent, then the consistency of T is not a theorem of the finitary metamathematics of T.

However, since, by the presumed classical soundness of finitary reasoning,[3] we can also say that

(7)   if T is inconsistent, then the consistency of T is not a theorem of the finitary metamathematics of T,

it follows (from (6) and (7)) that

(8)   the consistency of T is not provable in the finitary metamathematics of T.

But, from the definitions of consistency and real-soundness, we know that

(9)   if the consistency of T is not finitarily provable, then neither is the real-soundness of T.

And so it follows that

(10)   the real-soundness of T is not finitarily provable.

Furthermore, by (10) and the standard assumption on T (viz., that it is a formal theory containing elementary number theory), it follows that

(11)   there is no formal theory containing elementary number theory whose real-soundness is finitarily provable.

Hence, if to (11) we add the claim that

(12)   if there is no formal theory containing elementary number theory whose real-soundness is finitarily provable, then Hilbert's Program cannot be carried out for any appreciable body of ideal mathematics,

we arrive at our ultimate conclusion; namely, that

(13)   Hilbert's Program cannot be carried out for any appreciable body of ideal mathematics.

This, then, is what we take to be the Standard Argument against Hilbert's Program. Our task in succeeding chapters shall be that of demonstrating what we take to be some fatal problems concerning it. Our task in the remainder of this chapter is to give a preliminary statement of those problems. But before commencing either of these tasks, it seems advisable to issue a general statement concerning the overall character of our criticism, and who we think stands to be affected by it. Let me state at the very outset that none of my criticisms of the SA are criticisms to the effect that it fails to constitute a *finitarily correct* argument. Of course, it is *not* finitarily correct (witness, for example, premise (3) and the inference from (6) and (7) to (8)). But this should not be taken as a criticism of it. For it need only set forth a *cogent* case against Hilbert's Program. And if that case is based upon some classical or otherwise non-finitary reasoning, then so be it. The fact that Hilbert demanded finitary evidence to serve as the basis for his metamathematical defense of classical mathematical reasoning does not mean that only a finitary proof that such evidence doesn't exist could be used to dissuade him or any other rational being from his program. Indeed, it seems clear that even strong *empirical* evidence could, at least in principle, be used to demonstrate that his program cannot be carried out.

So, our claim that the SA does not refute Hilbert's Program is not to be taken as a consequence of the fact that it fails to give a *finitary* refutation of it. Accordingly, my criticisms of the SA will consist in charges to the effect that there is no good reason *of any kind* to accept the SA.

With this preliminary clarification in force, let us now proceed to the remaining task of this chapter; which is to give a statement of the major weaknesses of the SA.

3. THE STABILITY PROBLEM

The step from (3) to (4) in the SA is invalid. Basically, this is so because (3) does not guarantee that *every* formula of T that expresses a theorem of the finitary metamathematics of T will be provable in T, but only that *some* such formula will be. Hence, even if it is assumed that $Con_G(T)$ is not provable in T and that $Con_G(T)$ expresses a theorem of the finitary metamathematics of T, (3) need not be denied, since there might still be some formula other than $Con_G(T)$, expressing the same proposition that $Con_G(T)$ expresses, that *is* provable in T.

In order to repair the inference from (3) to (4), we thus need some principle to insure that any formula expressing the same proposition as $Con_G(T)$ will be provable in T only if $Con_G(T)$ is. And, practically speaking, it seems that the only knowledge upon which such a principle could be founded is knowledge that the properties of $Con_G(T)$ which render it unprovable in T are among the properties that any formula must have if it is to express the consistency of T. In other words, the gap separating (3) and (4) can be bridged only if it can be shown that unprovability-in-T is a stable or invariant feature of formulae of T possessing the wherewithal to express T's consistency. The Stability Problem is, then, precisely this: to show that every set of properties sufficient to make a formula of T a fit expression of T's consistency is also sufficient to make that formula unprovable in T (if T is consistent).

This problem is prompted by the possibility that the properties of $Con_G(T)$ which Gödel's proof calls upon to show the unprovability-in-T of $Con_G(T)$ may not all be included among those properties of $Con_G(T)$ which cause us to say that it expresses the consistency of T. Were this the case, the unprovability-in-T of $Con_G(T)$ and its expression of T's consistency would best be taken as sheer coincidence. And under such circumstances, we would not be at all inclined to cite $Con_G(T)$'s unprovability-in-T as evidence for the finitary unprovability of T's consistency.

In order to overcome this difficulty, the defender of the SA must find some strengthening of G2 that guarantees the unprovability-in-T of *every* formula which can reasonably be said to express the consistency of T. That is, he must isolate a set $\mathscr{C}$ of conditions on formulae of T such that every formula of T that can reasonably be taken to express the consistency of T satisfies $\mathscr{C}$, and the following generalization of G2 holds.

(Gen-G2) If T is consistent and $Con(T)$ is a formula of T which satisfies $\mathscr{C}$, then $Con(T)$ is not provable in T.

Now there are, to be sure, generalizations of G2 having the general form of Gen-G2, and we shall discuss the better-known ones in the next chapter. But none of these has been accompanied by a convincing argument to the effect that the generalizing conditions $\mathscr{C}$ *really are* conditions which must be required of formulae capable of expressing the consistency of T. Indeed, for the most part, those writing on Hilbert's Program seem to have been oblivious to the need to do so. In our view, this constitutes a serious deficiency in the literature on Hilbert's Program, since there are (as we shall argue in the next chapter) strong challenges to the reasonableness of such conditions that can be raised.

Yet despite its general neglect, there are still a few thinkers who have displayed a certain sensitivity to the Stability Problem. Chief among these (in our opinion) is Mostowski, who offers the following remark.[4]

If we assume that every combinatorial proof can be formalized within arithmetic, then Gödel's second theorem shows that Hilbert's program of proving in a purely combinatorial way the consistency of arithmetic is not realizable. This assumption is open for discussion, however, as we shall see later when we discuss Gentzen's theorem. *Another objection which can be raised against such interpretation of Gödel's second undecidability theorem is this: There are many formulae F strongly representing Z [i.e., T's proof relation] in T; Gödel's theorem is valid only for some such formulae. It is not immediately obvious why the theorem proved for just this formula should have a philosophical importance while a similar theorem obtained by a different choice of a formula strongly representing the same set Z is simply false.* (Mostowski [1966], p. 23, brackets and emphasis mine)

Still, even those who, like Mostowski, acknowledge the existence of the Stability Problem, appear to believe that its resolution requires only rather routine and straightforward reflection on the generalizing conditions that have gained currency. Since we do not agree with this, a goodly portion of our treatment of the Stability Problem is given to criticizing extant defenses of generalizing conditions. However, our attempt to vitalize the Stability Problem does not stop there. For, in addition to arguing that the currently existing defenses of the generalizing conditions are inadequate, we shall also present reasons for believing that no defense of these conditions *can be* successful.

4. STRICT INSTRUMENTALISM

The origins of this latter, more aggressive, argument go all the way back to the basic outlook in which Hilbertian instrumentalism is ultimately rooted. That outlook is one of serious respect for the practical limitations (e.g., limitations on the expenditure of time and energy, limitations on our ability to accommodate complexity, etc.) which govern all actual human epistemic activity. For the Hilbertian instrumentalist, the basic truths about human epistemic agents include such facts as that they are beings with strictly limited amounts of time and energy to expend in the pursuit of epistemic goals and that there are bounds on the complexity of epistemic acquisition devices from whose use they can derive any benefit.[5] As beings suffering such limitations, we are naturally concerned that our methods of epistemic acquisition be maximally efficient; i.e., that no alternative methods available to us yield a higher return on the expenditure of our limited epistemic resources than the ones that we have adopted. The Hilbertian defense of the ideal methods of classical mathematics, as we have presented it, is a proposal in this general spirit.

Yet the very limitations on human epistemic resources which attract the Hilbertian to the ideal method also pose clear limits to its utility. Thus, only ideal proofs falling below a certain level of length and/or complexity will be of any human utility. Ideal proofs exceeding that level will be of no value as devices of epistemic acquisition. Hence, the question of their reliability is of no concern to the Hilbertian instrumentalist.[6] Put more plainly, the Hilbertian instrumentalist is obliged to establish the reliability only of such ideal proofs as are of feasible length and complexity. And while this may strike the reader as a rather pedestrian observation, we shall see in succeeding chapters that it is the source of many a serious problem for the Gödelian challenge.

These problems stem from a single consequence of the above limitation of the Hilbertian's responsibilities which we now state.

> (*The Thesis of Strict Instrumentalism*): Of the infinitely many ideal proofs constructible in a given system T of ideal mathematics, only finitely many of them are of any value as instruments of human epistemic acquisition.

This thesis (which, hereinafter, we shall refer to as the TSI) can be argued for in a variety of different ways. But we shall base our case for it on some very straightforward length-of-proof considerations.

When we speak of the length of an ideal proof of the system T we shall mean the number of occurrences of characters of T's alphabet that appear in the proof. There are, of course, other measures of length-of-proof that might be used and that enjoy greater currency in the literature of proof theory (e.g., the number of logical operator occurrences, the number of quantifier alternations occurring in the line of the proof containing the maximum number of quantifier alternations of all lines in the proof, etc.). But our simple measure provides us with a clear and relatively uncontroversial argument for the TSI, and this is our reason for employing it.

So, we claim that there is a finite bound on the number of character-occurrences that can appear in a humanly useful ideal proof. At least this is so if our model for the epistemic application of ideal proofs is that which is offered by the Hilbertian; namely, the Metamathematical Replacement Strategy. For, according to the Replacement Strategy, the epistemic application of a given ideal proof is via its being incorporated into an M-proof. However, the process of incorporation is not one wherein the ideal proof becomes a subproof of the M-proof. Rather, as was stressed repeatedly in our earlier discussion of Frege's Problem, incorporation is via metamathematical evaluation. And this metamathematical evaluation involves a determination that the ideal proof in question really is a proof in a certain formal system.

Yet determining that a given ideal proof *is* a proof in a certain formal system entails a character-by-character surveyal of it. And since it is true both that the surveyal of each individual character requires some finite expenditure of time and effort (even if only a very small one) and that there is a finite bound on the amount of time and effort that it is humanly feasible or worthwhile to expend in the construction of an M-proof, it follows that there is a finite bound on the number of character-occurrences that can appear in a humanly useable ideal proof. From this, in turn, the TSI clearly follows.

With the TSI at our disposal, it is easy to show that the Hilbertian instrumentalist is, strictly speaking, only obliged to establish the real-soundness of *a finite number* of the ideal proofs constructible in a given system. For surely, since he only *advocates* using such ideal proofs as are humanly feasible, he is only obliged to *defend* the reliability of those that are. Hence, since (by the TSI) only a finite set of the ideal proofs constructible in a given system are humanly useful, only the reliability of the proofs in that set need be established. This, as we shall see in the next section, causes serious problems for the SA.

5. THE CONVERGENCE PROBLEM AND THE PROBLEM OF STRICT INSTRUMENTALISM

The most serious problems concern premises (3) and (12). The problem which arises for (3) is this: in light of the TSI, it is no longer clear that the content of the Hilbertian's ideal mathematics (conceived now as consisting essentially in the *feasible* body of ideal proofs of a given system) and the content of finitary reasoning will "converge" in the manner demanded by (3). That is to say, there is no longer any reason to suppose that the set of *feasibly* provable real theorems of the ideal system T will subsume the set of finitarily provable formulae of T.

Nor is there any reason to suppose that *every* finitarily provable formula of T is more efficiently proven by an ideal proof in T than by a purely finitary proof; ideal methods cannot always be expected to be more efficient than their contentual counterparts.

But since the Hilbertian only advocates the use of such ideal proofs as would yield a gain in efficiency when used in lieu of their contentual counterparts, it follows that (strictly speaking) it is only such proofs whose real-soundness he must establish. Thus, we have a second reason for doubting that the ideal methods to which the Hilbertian is committed theorem-wise subsume finitary reasoning.

This, briefly stated, is the Convergence Problem. In focusing our attention on premise (3) of the SA, it reminds us of certain other attempts (e.g., those of Ackermann [1940] and Gentzen [1936]) to rescue Hilbert's Program from the threat posed by G2. Yet, in spirit, it is radically different from these other proposals. For they all seek to put pressure on (3) by arguing for an *extension of* the (original) conception of *finitary evidence* which would include methods going beyond the bounds of what is formalizable in ideal number theory. We, on the other hand, are arguing for a *restriction of* the usual conception of *the ideal method* that would bring it into closer agreement with the dictates of the TSI. In so doing, we bring considerable pressure to bear on (3) without having to argue for a liberalization of the conception of finitary evidence. But still more importantly, we focus attention upon certain central features of Hilbert's instrumentalist outlook (viz., its concern with the *feasibility* and *efficiency* of ideal proofs) that have been neglected by previous assessments of it, and yet which are critically important to its proper evaluation.

A parallel application of the TSI brings like pressure to bear on premise (12). Since, on the view of the Hilbertian instrumentalist, the appreciability of a given body of ideal mathematics is determined by its utility as a device for promoting epistemic acquisition of the truths of real mathematics, it follows from the TSI that, for every system T of ideal mathematics, there is some finite subset of T's proofs which is just as appreciable as the entire infinite set of T's proofs.[7] Hence, given the TSI, we are led to conclude that T is no more appreciable a body of ideal mathematics than are various of its finite subsystems.

With this conclusion at our disposal it is possible to argue for the falsehood of (12). For "elementary number theory", as conceived of in that premise designates an infinite system of proofs. And that being so, it cannot be contained in any of the finite systems of ideal proofs which, as we have just concluded, are equi-utilitous with the infinite systems of ideal proofs containing them. Hence, assuming that at least some formal system T is an appreciable system of ideal proofs, it can be shown that not every appreciable system of ideal proofs contains elementary number theory. Thus, our case against (12) is made.

Here, however, we must warn the reader not to confuse this argument with what might look like a more positive defense of Hilbert's Program. We are not suggesting that the Hilbertian's quest for a finitary defense of the ideal method is, in principle, accomplished; that all he need now do is to go through the useful ideal proofs of T one by one and show of each that its finitary conclusion is also finitarily provable. (Though, to be sure, if the finitely many useful ideal proofs of T are real-sound, then, in principle, one can give a finitary proof of each real theorem thus proven. And these finitary proofs, taken collectively, would serve as a finitary proof of the reliability of the instrumentally useful portion of T.) For even granting that the Hilbertian instrumentalist is only obliged to establish the real-soundness of T's useful portion, it does not follow that the Hilbertian's program can be carried out by means of the case-by-case strategy just described.

This is so because the Hilbertian actually requires *more* than a finitary proof of the reliability of those ideal methods whose use he advocates (i.e., the *useful* ideal methods). He requires, in addition, that the proof of reliability be incorporated into a scheme of M-proofs which provides for more efficient epistemic acquisition of a certain class of real mathematical propositions than does any scheme of O-proofs; in short, he requires that the replacement of O-proofs by M-proofs carry with it a gain in epistemic efficiency. And it is quite clear that the case-by-case strategy of proving the soundness of T's useful ideal proofs produces no such gain in efficiency. Indeed, it actually represents a task that is lengthier, more complex, and hence more demanding of time and effort, than the task of simply constructing the corresponding set of O-proofs, since, on the case-by-case strategy, the soundness proof upon which the M-proofs are to be based actually requires the construction of the very O-proofs whose construction the Hilbertian seeks to avoid (via appeal to his Replacement Strategy).

Adopting the case-by-case approach to proving soundness would, therefore, force the Hilbertian instrumentalist to renounce his central amd motivating claim: namely, that the ideal method, under the mechanism prescribed in the Replacement Strategy, possesses epistemic utility by making the epistemic acquisition of real mathematics more efficient.[8] Hence, the case-by-case approach to soundness is unacceptable to the Hilbertian *even if* it provides for a finitary proof of the soundness of his ideal proofs.

Does this, in effect, overturn our objection to premise (12) and thus blunt the edge of the Problem of Strict Instrumentalism (which is our name for that objection)? It may appear to. For the only visible alternative to the case-by-case approach is that of embedding the useful proofs of T in some inductively defined set of proofs. (One would then show directly that the proofs cited in the basis clause of the definition are all real-sound, and show also that real-soundness is preserved by the rules of construction figuring in the induction clause of the definition.[9]) But then the question is this: "Is there any inductively defined set of proofs of T, *other than T itself*, in which the entire set of useful ideal proofs of T can be embedded?"[10] If the answer to this question is negative, then it would appear that the only *feasible* way to prove the soundness of T's useful part is to prove the soundness of T itself. And if this is so, our case against premise (12) of the SA is nullified, and there may indeed be no suitable way of proving the soundness of T's useful proofs.

There is, however, good reason to believe that the useful part of T can be embedded in an inductively defined set of proofs other than T itself. Indeed, for at least some very important cases of T,[11] it appears possible to embed the set of T's useful proofs in an inductively defined system of proofs whose consistency is provable in T! We call this system the Hilbertian residue of T, and designate it as 'T_H'.

We shall discuss the significance of Hilbertian residues more fully in Chapter V. Here we shall only sketch the general principles guiding their construction by saying how, in idealized terms, the useful portion of T is to be determined. The basic idea is that if, beginning with T, we successively eliminate from it

(i)   all ideal proofs of real formulae that are too long or complex to be of any human epistemic utility

and

(ii)   all ideal proofs of real formulae that have an equally short and simple real proof

and, finally,

(iii)   all real proofs of real formulae,[12]

then we end up with a set of proofs of T that will include all of the humanly useful ideal proofs of T.[13]

Following these eliminations, we take the set of axioms of T appearing in the remaining proofs and close it under the logic of T. Of course this closure will introduce into T_H some of the proofs that were eliminated by (i)—(iii). But it will still eliminate from T_H those proofs of T containing axioms which do not appear in any instrumentally useful proof of T. (And, at least for non-finitely axiomatizable cases of T (e.g., PA and ZF), the axioms thus eliminated comprise the major portion of the axioms of T.) In order to see that this is so, consider the effect of clause (i). In conjunction with the TSI, it implies that the set of proofs from which the axioms of T_H may be selected is finite.[14] Hence, since the number of axioms of T appearing in any one of these proofs is also finite, it follows that the set of axioms of T_H will be finite. The finite axiomatizability of T_H, as we shall see in Chapter V, is one of its most important features and one which enables us to pose a stiff challenge to the SA and any other argument based on a like application of G2.

The immediate significance of the invention of Hilbertian residues, then, is this: they provide for the embedding of the instrumentally useful ideal proofs of T in an inductively defined set of proofs of T which is smaller than T itself. As such, they offer the Hilbertian the possibility of proving the soundness of T's useful portion without having to prove the soundness of T. And, at the same time, they enable him to avoid the disastrously inefficient case-by-case approach to soundness. Thus, despite the Gödelian challenge, it may still prove possible to give both a *finitary* and a *feasible* demonstration of the soundness of the useful ideal methods.[15] This at any rate, is the hope that is held out by the introduction of Hilbertian residues.

6. CONCLUSION

Let us close by orienting ourselves a bit. The Gödelian challenge (i.e., the SA), as we see it, is fundamentally an attempt to show that the Dilution Problem cannot be satisfactorily resolved.[16] For the central thrust of the Gödelian challenge is that there is no finitary proof of the soundness of the ideal methods, and this, as we have already seen, raises the specter of dilution. If the soundness of the ideal methods cannot be proven finitarily, then the M-proofs with which the Hilbertian plans to replace the problematic O-proofs will not be finitary. Hence, given the special strength of finitary proof (i.e., that expressed in the Principle of Weak Optimality), such replacement then threatens dilution, or at least reduction of the range of O-proofs that can be replaced by M-proofs without dilution.

As objections to the Gödelian challenge, therefore, the Stability Problem, the Convergence Problem, and the Problem of Strict Instrumentalism may be seen as attempts to counter the Gödelian's case against the solvability of the Dilution Problem. And if my assessment of the situation is at all correct, the argument of the following two chapters will show that the prospects for a finitary soundness proof and, hence, for a satisfactory resolution of the Dilution Problem, are not nearly so poor as the Gödelian challenge would have us believe.

NOTES

[1] In the 1931 paper Gödel shied away both from this claim and from any suggestion that his theorem refuted Hilbert's Program. He wrote that he wished

. . . to note expressly that Theorem XI [G2] (and the corresponding results for M and A) do not contradict Hilbert's formalistic viewpoint. For this viewpoint presupposes only the existence of a consistency proof in which nothing but finitary means of proof is used, and it is conceivable that there exist finitary proofs that cannot be expressed in the formalism of P (or of M or A)." (Gödel [1931], p. 615, square brackets mine)

However others, most notably Bernays (cf. Bernays [1935a, 1939, and 1941]), have argued that Gödel's caution is unnecessary, and eventually Gödel himself was won over to their viewpoint (cf. Gödel [1958], p. 133). Today it represents the nearly universal opinion of philosophers of mathematics and logicians.

[2] The inference here may be made slightly more perspicuous by recasting (4) in equivalent but different terminology to get

(4 =) if $Con_G(T)$ is not provable in T, then that statement which is expressed by $Con_G(T)$ is not a theorem of the finitary metamathematics of T.

[3] By the classical soundness of finitary reasoning we mean this: every finitary theorem (= finitary truth) is, when interpreted classically, true.

[4] Another clear exception is Resnik [1974]. Kreisel too shows some sensitivity to the point in [1958, 1971, 1976] and in his joint pieces with Takeuti [1974], and Levy [1968]. Finally, there are allusions to such a problem in Feferman [1960], pp. 36–40.

[5] I believe that some such practical concerns must be part of what lies behind *any* sensible version of instrumentalism.

[6] And this is so even if the infeasible ideal proof is more nearly feasible than any of its contentual counterparts. Gains or losses in efficiency in the realm of the infeasible are simply irrelevant to the human epistemic utility and, hence, the instrumentalistic utility, of the ideal method.

[7] Here it may look as if we are assuming that, for purposes of discussing utility, a formal system may be identified with its set of derivations. But actually we're not. We would allow that part of what determines the identity of a formal system is the *process* by which the derivations are generated. So, we would identify T's utility with that of its derivations *as generated by a certain process*. And we interpret the TSI as saying that the epistemic utility of an infinite set of derivations taken as generated by a given process is identical to that of certain of its finite subsets taken as generated by the *same* process.

[8] It should also be remembered that the mechanism of the Replacement Strategy is key to the Hilbertian since it provides the basis for his response to Frege's Problem.

[9] Of course, the useful proofs of T are assumed to be "embedded" in this inductively defined set in such a way as to avoid the need to prove their soundness case-by-case. Hence, it is assumed that they do not all, or even

mostly, fall under the basis clause of the definition. Without such an assumption, there is no reason to suppose that the inductive approach to proving soundness is any less laborious and inefficient than the case-by-case approach.

[10] I am treating the formal system T here as if it is an inductively defined set of proofs. There are, of course, other ways of thinking of formal systems (e.g., as merely sets of theorems), but this particular conception suits present purposes best.

[11] PA and ZF, for example.

[12] Clearly, any proof of real-soundness such as that envisaged by Hilbert presupposes that some sort of formalization (not necessarily complete) of finitary reasoning can be given. For the epsilon-elimination strategy is designed to syntactically transfrom an *ideal* formal derivation into a *real* formal derivation of the same result. Thus, it supposes that there is a syntactical form that a derivation can have that definitely qualifies it as a *real* derivation (i.e., a formal expression of a real proof). See the references to the elementary calculus and the elementary calculus with free variables in Leisenring [1969] and Kneebone [1963] for more on this subject.

[13] Let us add at this point a word about motivation. Clause (i) is motivated by the fact that the Hilbertian instrumentalist is committed, in principle, to defending only such ideal proofs as stand to be of some instrumental epistemic value to their users. Clause (ii) has as its rationale the fact (already discussed in Chapters I and II) that the Hilbertian instrumentalist takes some O-proofs (viz. those which, in Chapter I, we referred to as the unproblematic real proofs) to be incapable of gainful replacement by M-proofs. Finally, clause (iii) is prompted by the fact that the Hilbertian instrumentalist is only obliged to prove the real-soundness of ideal proofs and not of real proofs.

[14] We are tacitly assuming that the alphabet of T is finite. Any of the (reflexive) theories that we are interested in (e.g., PA or ZF) can be (and typically are) formulated in such a way as to satisfy this constraint.

[15] On our view, the feasibility of the soundness proof is quite as vital to the Hilbertian as its finitariness. That is why the case-by-case approach to soundness is unacceptable even if it yields a finitary proof. The importance of feasibility has been overlooked in the literature on Hilbert's Program, and so we hope that the present discussion has helped the reader to see that the roots of the feasibility constraint penetrate even more deeply into the Hilbertian's ideology than do those of the finitariness constraint. It is at least conceivable that Hilbertian instrumentalism be based upon a non-finitary soundness proof, since whatever dilution resulted from this might be superceded by the gains made in the expansion of the noetic corpus. But it is wholly inconceivable that Hilbertian instrumentalism be based on an infeasible soundness proof. For then the M-proofs would be of no use as substitutes for O-proofs and the whole Replacement Strategy, which is the very heart of Hilbert's instrumentalism, would collapse.

[16] To be more specific, the Dilution Problem lies behind premises (11) and (12) since part (though only part) of what motivates those premises is the doctrine that the only adequate proof of soundness is a finitary one.

CHAPTER IV

THE STABILITY PROBLEM

1. INTRODUCTION

Presumably, if the Gödelian is to find a solution to the Stability Problem for a given system T (T being an ideal system whose soundness is in question, and therefore a system whose syntax is to be represented or "arithmetized") he must locate a set $\mathscr{C}$ of conditions on formulae of T (T now being treated also as the system *in which* the syntax of T is to be represented) such that (1) every formula of T that can reasonably be said to express the consistency of T satisfies the conditions in $\mathscr{C}$, and (2) no formula of T that satisfies $\mathscr{C}$ can be proven in T provided that T is consistent. This being so, the Gödelian's success in dealing with the Stability Problem will evidently depend crucially upon his ability to defend the reasonableness of his choice of $\mathscr{C}$.

The chief business of this chapter is, therefore, first to develop and thence to critically evaluate the Gödelian's choice(s) of $\mathscr{C}$. Overall, our findings are negative. We find no satisfactory defense of the reasonableness of the actual historical choices of $\mathscr{C}$; nor do we find any sufficient grounds for believing that one is in the offing. Our conclusion, therefore, is that the Gödelian has made but little progress towards finding a solution to the Stability Problem.

The development of our argument proceeds in three main stages. In the first, we present (and develop some of the background of) the traditional choice of $\mathscr{C}$. In the second, we make a very brief attempt to characterize the general sense in which an arithmetization of T's syntax is supposed to serve as a representation of it. And in the third, we attempt to assess the reasonableness of the representational scheme which results when arithmetization is constrained by the usual choice of $\mathscr{C}$. With this

93

general plan in mind, let us now turn to the discussion of the first stage.

2. THE STANDARD CHOICE OF $\mathscr{C}$

In 1939, Hilbert and Bernays set forth what, with a few modifications, was eventually to become the standard set of conditions for obtaining a generalized version of G2. Gödel's original proof of the second theorem was directed at the consistency formula based on the *particular* provability formula that he constructed in order to prove his first theorem. He observed that the reasoning used in the proof of the first theorem could be "formalized" in T (= any of the systems for which the first theorem was proven) itself, and that if it was, the second theorem would result.

This suggests an obvious strategy for generalizing G2; namely, isolate those features of the provability formula that are crucial to the proof of G1 (thus producing a generalized version of G1), and show that those features are all establishable in T. As we shall now try to show, this is essentially what the proof of the usual generalization of G2 does.

To begin with, then, we shall give the proof of the generalized version of G1.[1] It holds for any provability formula for T (briefly, 'Prov$_T(x)$') satisfying the following three properties.

(i)   The Diagonalization Property (DIAG, for short): there is some formula G of T such that $\vdash_T G \equiv \sim \text{Prov}_T(\ulcorner G \urcorner)$.[2]

(ii)   The Local Provability Completeness Property (LPC, for short): for every formula A of the language of T, if $\vdash_T A$, then $\vdash_T \text{Prov}_T(\ulcorner A \urcorner)$.

(iii)   The Local Provability Soundness Property (LPS, for short): for every formula A of the language of T, if $\vdash_T \text{Prov}_T(\ulcorner A \urcorner)$, then $\vdash_T A$.[3]

With these conditions at our disposal, the proof of our generalized version of G1 proceeds easily in two parts.

PART I: If T is consistent, then it is not the case that $\vdash_T {\sim} G$.

Proof. Suppose that $\vdash_T {\sim} G$. From this supposition and DIAG, we are led to infer that $\vdash_T \mathrm{Prov}_T(\ulcorner G \urcorner)$. And, by LPS, this in turn leads us to conclude that $\vdash_T G$. Thus, if $\vdash_T {\sim} G$, then T is inconsistent. Hence, by contraposition of this conditional, we get Part I. QED

PART II: If T is consistent, then it is not the case that $\vdash_T G$.

Proof. Suppose that $\vdash_T G$. This supposition leads first, by DIAG, to $\vdash_T {\sim} \mathrm{Prov}_T(\ulcorner G \urcorner)$, and secondly, by LPC, to $\vdash_T \mathrm{Prov}_T(\ulcorner G \urcorner)$. So, if $\vdash_T G$, then T is inconsistent. Thus, by contraposition of this conditional, we get Part II. QED

The generalization of G2 with which we shall be working does not require the above proof in its entirety, but rather only Part II. However, it makes use of the proof of Part II in two different ways. First, it calls for the "formalization" of Part II in T to obtain the result that

(Gen G2 $\triangle\triangle$) $\vdash_T \mathrm{Con}(T) \supset {\sim} \mathrm{Prov}_T(\ulcorner G \urcorner)$,[4]

which, when taken together with DIAG leads us in turn to the conclusion that

(Gen G2 $\triangle$) $\vdash_T \mathrm{Con}(T) \supset G$.

Secondly, it invokes Part II itself (in combination with (Gen G2 $\triangle$)) to get

(Gen G2)   If T is consistent, then it is not the case that $\vdash_T \mathrm{Con}(T)$,

which is the generalized version of G2 that we have been looking for.

It thus becomes clear that the conditions $\mathscr{C}$ that must be placed upon '$\mathrm{Prov}_T(x)$' in order to insure our generalized version of G2 include those that are required for the proof of Part II of our generalized version of G1, plus those that are needed to "formalize" this proof in T. Looking first to Part II, we see that its proof requires DIAG and LPC. So, we will need to include those conditions in $\mathscr{C}$. But the conditions required for the formalization in T of Part II are a little harder to extract. At first glance, one is inclined to say that since the proof of Part II itself requires DIAG and LPC, the formalization of Part II in T must require precisely the formalization of each of these conditions in T. That is, it must require that

(F-DIAG)   there is some formula G of the language of T such that $\vdash_T \mathrm{Prov}_T(\ulcorner G \equiv\ \sim \mathrm{Prov}_T(\ulcorner G \urcorner)\urcorner)$,

and that

(F-LPC)   for all formulae A of the language of T,
$\vdash_T \mathrm{Prov}_T(\ulcorner A \urcorner) \supset \mathrm{Prov}_T(\ulcorner \mathrm{Prov}_T(\ulcorner A \urcorner)\urcorner)$.

This estimate of the requirements of the formalization of Part II is, however, inaccurate in two respects. First, it treats F-DIAG as an added condition when, in fact, it follows from DIAG and LPC. And secondly, it fails to include a condition which is needed to codify that step in the proof of Part II where, from the hypothesis that $\vdash_T G$, we call upon DIAG to infer that $\vdash_T\ \sim \mathrm{Prov}_T(\ulcorner G \urcorner)$. More than DIAG is involved here, for it is also being assumed that *modus ponens* is an acceptable form of inference in T. And, if we are to codify this form of inference in T, we must require that '$\mathrm{Prov}_T(x)$' satisfy the following condition:

(F-MP)   for all formulae A, B of the language of T,
$\vdash_T (\mathrm{Prov}_T(\ulcorner A \supset B \urcorner) \wedge \mathrm{Prov}_T(\ulcorner A \urcorner)) \supset \mathrm{Prov}_T(\ulcorner B \urcorner)$.

Thus, the net addition that the formalization of Part II makes to $\mathscr{C}$ is F-LPC and F-MP. $\mathscr{C}$, therefore, is to be taken as comprised of these conditions: DIAG, LPC, F-LPC, and F-MP. Moreover, because the consistency formula (i.e., $\mathrm{Con}(T)$) is constructed in a simple and direct way from the formula expressing provability, we may properly think of these conditions (which we shall, hereafter, refer to collectively as the "Derivability Conditions") not only as governing our choice of formal expressions of provability, but also of our formal expressions of consistency.

The Derivability Conditions, then, constitute what we take to be the standard choice of constraints on consistency expressions needed to produce a suitable generalization of G2.[5] As such, they will serve as the focal point of the critical discussion of the Stability Problem that is to follow. But before proceeding with that discussion, I should just like to note that what has thus far been said about the Derivability Conditions is in no wise capable of *justifying* their use as constraints governing consistency expressions. Rather it only *explains why* the anti-Hilbertian (i.e., the advocate of the SA) must produce such a justification. The Derivability Conditions are required for the proof of Gen G2,[6] so if the anti-Hilbertian is to make his case, he must defend their propriety. Since the mere fact that they are required for the proof of Gen G2 does not constitute such a defense, more must be said on their behalf. Various proposals to this effect have been made in the literature, and their evaluation is that which shall, by and large, occupy our attention for the remainder of this chapter. However, we must first get a clearer picture of how (i.e., in what sense) it is that the whole enterprise of arithmetization (of which the Derivability Conditions are but a part) is supposed to contribute to the evaluation of Hilbert's Program.

3. ARITHMETIZATION

At its most basic level, arithmetization is supposed, by the advocate of the SA, to be a device which enables us to make inferences to what can be done in the informal metamathematics of a formal system T from discoveries about what can be done in T itself. More specifically, it is supposed to be a device which allows us to infer that the consistency of T cannot be proven in its informal finitary metamathematics from the hypothesis that a certain formula $Con(T)$ is not provable in T. This inference is supposedly sanctioned by the special "correlation" that arithmetization establishes between $Con(T)$ and the proposition that T is consistent.

What would appear to be the usual conjecture about the nature of this "correlation" and how it arises is that it is a *semantical* relationship between the formula $Con(T)$ and the (finitary) metamathematical proposition that T is consistent. It is this semantical relationship that is being talked about when it is said that Con(T) "expresses" the consistency of T.

However, this relationship of "expression" is more complicated than the usual semantical relationship which links a sentence of a formal language with the proposition that is its standard interpretation. For the standard interpretation of Con(T) is not the proposition that T is consistent, but rather a certain (rather complicated) arithmetic proposition. In order to pass from this arithmetic proposition to the metamathematical proposition that T is consistent, an additional step of "interpretation" is necessary; namely, that in which the inverse of the Gödel numbering operation is applied in order to metamathematically decode the arithmetic proposition which is Con(T)'s standard interpretation.

Thus, the semantical relationship between Con(T) and the proposition that T is consistent is made up of two different components. One is the usual semantical relationship by which the formula Con(T) is connected to the arithmetic proposition that is its interpretation under the standard semantics for T. And the other is that which Gödel numbering induces between the genuine arithmetic proposition which is Con(T)'s standard interpretation, and the genuine metamathematical proposition which asserts T's consistency.

With this bipartite view of the mechanism of arithmetization in place, we can now attempt to say how it is that arithmetization is supposed to provide for the inference of the finitary unprovability of T's consistency from Con(T)'s unprovability in T. So far as we can tell, that inference is conceived of as follows:

(P1)   Con(T) is not a theorem of T.

(P2)   Con(T) expresses the proposition that T is consistent.

(P3)   All finitarily provable propositions of the informal metamathematics of T are expressed by formulae that are theorems of T.

(C)   The proposition that T is consistent is not finitarily provable.[7]

The Stability Problem challenges this inference without contesting any of its premises. In other words, it challenges the *validity* of this inference. And the basis for such a challenge resides in the simple fact that the proposition that T is consistent might be expressed by formulae other than $\text{Con}(T)$. This being so, there is no inconsistency in accepting P1–P3 of the above inference while yet denying its conclusion. Therefore, the usual account of the inference from $\text{Con}(T)$'s unprovability in T to the finitary unprovability of T's consistency fails to justify it.

One possible diagnosis of what is wrong with that account would put the blame on its appeal to a semantical connection between $\text{Con}(T)$ and the proposition that T is consistent (which proposition we shall hereafter refer to as '$\text{CONSIS}(T)$').[8] The invalidity of the inference in question is, after all, due precisely to the fact that the semantical relationship that exists between $\text{CONSIS}(T)$ and $\text{Con}(T)$ might also exist between $\text{CONSIS}(T)$ and formulae other than $\text{Con}(T)$. Therefore, since the semantical relationship between $\text{CONSIS}(T)$ and $\text{Con}(T)$ is (by itself) incapable of sustaining the inference, perhaps the basic mistake lies in ever bringing it into the picture in the first place.

Such a proposal is not entirely absurd since it is at least conceivable that an inferential tie between $\text{Con}(T)$'s unprovability in T and the finitary unprovability of $\text{CONSIS}(T)$ be based on something other than a semantical connection between the two. There are, after all, principles like LPC which, though they provide for an inferential connection between the arithmetized and the unarithmetized metamathematics of T, do not rely upon any semantical relationship between the formula '$\text{Prov}_T(\ulcorner A \urcorner)$' and the metamathematical proposition that A is provable in T.[9]

Still, the idea of trying to preserve the inferential tie between $\text{Con}(T)$'s unprovability in T and the finitary unprovability of $\text{CONSIS}(T)$ without appealing to a relation of semantical expression between $\text{Con}(T)$ and $\text{CONSIS}(T)$ is not a very attractive one. For even if it is in principle possible that this be done, it is nonetheless the case that we haven't the slightest idea of how to go about it. A strict parallel to LPC (i.e., a condition asserting that if not-$\vdash_T A$, then $\vdash_T \sim \text{Prov}_T(\ulcorner A \urcorner)$) is simply out of the question, since it is demonstrably false.[10] And more plausible principles which avoid the charge of demonstrable falsehood would all seem to depend upon a claim of semantical relatedness between the formula ' $\sim \text{Prov}_T(\ulcorner A \urcorner)$ ' and the proposition that A is not provable in T to retain their plausibility.

We believe, therefore, that the inference from $\text{Con}(T)$'s unprovability in T to the finitary unprovability of $\text{CONSIS}(T)$ (and, hence, the anti-Hilbertian application of arithmetization) is critically dependent upon the claim that $\text{Con}(T)$ semantically expresses $\text{CONSIS}(T)$. And because of this, we see only one way for the anti-Hilbertian (i.e., the advocate of the SA) to deal with the Stability Problem; namely, to argue that *all* formulae (of T) expressing $\text{CONSIS}(T)$ are like $\text{Con}(T)$ in being unprovable in T. With such an argument at his disposal, the anti-Hilbertian could replace P1 and P2 of the above inference with the single premise that

(P1−P2)*   no formula of T that expresses $\text{CONSIS}(T)$ is provable in T,

and thus facilitate the inference from the unprovability in T of any given formula expressing $\text{CONSIS}(T)$, to the conclusion that $\text{CONSIS}(T)$ is not finitarily provable. That way, the inference upon which the anti-Hilbertian so critically depends would achieve respectability.

But how should the anti-Hilbertian go about arguing that all formulae of T that express $\text{CONSIS}(T)$ are like $\text{Con}(T)$ in being unprovable in T? One straightforward strategy would be to simply examine all of the particular formulae of T thought to express the consistency of T and inductively infer the unprovability in T of all formulae expressing $\text{CONSIS}(T)$ from the unprovability in T of all members of the class examined.

Such an argument would surely provide some statistical evidence for the claim that all formulae expressing $\text{CONSIS}(T)$ are unprovable in T. But the various attempts to get a generalized version of G2 (e.g., those by Bernays, Löb, Feferman, and Jeroslow) have surely had something quite different in mind. Such work is intended to reveal a necessary rather than a merely statistical connection between a formula's ability to express CONSIS(T) and its unprovability in T. And it sets about doing so by presenting some set of conditions on Con(T) which we are supposed to recognize as being at once, both *necessary* for Con(T)'s ability to express CONSIS(T), and *sufficient* to guarantee its unprovability in T.

The position that we shall be arguing for in the remainder of this chapter is that there is no analysis of what is required of a formula capable of expressing CONSIS(T) that would in any way implicate the Derivability Conditions. Therefore, we are of the opinion that nothing thus far produced gives any promise of yielding an alternative to the purely statistical or inductive approach to the Stability Problem.

We believe, moreover, that there is reason to doubt the adequancy of even the inductive or statistical approach. This is owing to the fact that it is possible to argue against the necessity of the condition F-LPC. To the extent that this argument is successful, it suggests that satisfaction of F-LPC is a purely accidental feature of some formulae expressing CONSIS(T) and, therefore, that there is a limit to how far any statistical correlation between formulae expressing CONSIS(T) and formulae satisfying the Derivability Conditions can be taken as indicative of what *must* be the case.[11]

Overall, then, we are sceptical of the chances of any adequate resolution of the Stability Problem arising from current attempts to generalize G2. And hopefully the grounds for our scepticism will become clearer and more convincing as we examine the various defenses of the Derivability Conditions that exist in the literature.

4. MOSTOWSKI'S PROPOSAL

In his very useful survey, *Thirty Years of Foundational Studies*, Andrzej Mostowski offers a view of what a good arithmetization of the metamathematics of T ought to do. This view, which he takes to be (to one extent or another) tacit in the work of Gödel

[1931], Hilbert and Bernays [1939], and Feferman [1960], is that the best T-theoretic representation of a metamathematical notion M is provided by that formula of T that does the best (i.e., the most complete) job of registering the intuitive truths regarding M as theorems of T. He states this view, modulo an assumed Gödel numbering of T's syntax, in the following way.

> The general problem . . . can be described as follows: there is given, on the one hand, a set X of integers (or of pairs, triples, etc.) and, on the other hand, a formal language. We are looking for the best possible definition of X in T, i.e., *for a definition which makes, of all the intuitively true formulae involving X, as many as possible provable in T*.[12] (Mostowski [1966], p. 25; emphasis mine)

Mostowski "rates" different candidates for representation according to such a standard. Gödel's condition (in his first theorem) is that we make the representing formula of ψ_T ($=$ the set of number-pairs $\langle x, y \rangle$ such that x is the Gödel number of a derivation in T of the formula whose Gödel number is y) *strongly represent* ψ_T.[13] This condition forces truths of the form "A is a proof-in-T of B" and truths of the form "A is not a proof-in-T of B" to be registered as theorems of T. And since the sum total of all truths of these forms captures the extension of the proof-of relation for T, Gödel's condition might be said to enforce the extensional adequacy of our representation of this notion.

But, Mostowski notes, not *all* truths concerning the proof-of relation are particular truths (i.e., truths either of the form "A is a proof-in-T of B" or of the form "A is not a proof-in-T of B"). There are also general truths to be considered. And it is *not* the case that all formulae strongly representing the relation "X is a proof-in-T of Y" do an equally good job of registering the general truths regarding this notion as theorems of T. Hence, when judged according to the basic view of adequate representation suggested by Mostowski, the condition of strong representation turns out to be insufficiently discriminating.

Mostowski then goes on to credit first Bernays and then Feferman with showing how the proof (and, indeed, the truth) of a generalized version of G2 depends upon the adoption of a more sensitive and refined standard of arithmetic representation;

namely, one which attends to the registration (as theorems of T) of general as well as particular truths concerning the notion that is to be represented. The extensional standard, in Mostowski's opinion, correctly identifies the root mechanism of arithmetization as consisting in the formal "mimicking" of truths regarding the notion to be represented. But it errs in suggesting that the only truths which matter are those which, taken collectively, give the extension of the notion that is to be represented. The moral of the analyses of Bernays and Feferman, according to Mostowski, is that some metamathematical tasks (e.g., the evaluation of Hilbert's Program) call for a degree of fidelity between an arithmetical representation and the notion it represents which exceeds that which an extensionally adequate representation can guarantee. The degree of fidelity between a metamathematical notion M and the formula $\mathscr{M}$ of T which, under arithmetization, is to represent it, is determined by the proportion of truths regarding M which are registered as theorems of T when M is "translated" as $\mathscr{M}$. But, we must be careful to take general as well as specific truths, regarding M into the mix.

Such, at any rate, is the view of arithmetization which Mostowski urges. And it implies, or at least suggests, a corresponding defense of the Derivability Conditions. That defense, as we see it, would proceed roughly as follows: if we take as our T-theoretic representation of the notion of provability-in-T a formula satisfying the Derivability Conditions, then we will codify more truths concerning that notion as theorems of T, and hence do a better job of representing it, than we would otherwise do.

However, the cogency of this reasoning is convincingly called into question by an observation which, curiously enough, Mostowski himself makes; namely, that regardless of which formula we choose to represent the notion of provability-in-T (or such cognate notions as X's being a proof in T of Y), not all truths involving that notion will be codifiable as theorems of T.[14] In our opinion, such an observation raises a very serious problem for Mostowski: namely, that if, regardless of which formula we choose to represent the notion of proof-in-T, there will always be truths involving it that cannot be codified as theorems of T, then what point is there to registering as theorems of T as many truths concerning that notion as is possible? This problem is sharpened by the realization that either representation of the notion of proof-in-T requires the codification as a theorem of T of every truth concerning that notion, or it does not. If the former, then the proper conclusion to draw from Mostowski's observation (and also from the likes of G1 and G2) is simply that there is no good representation of the notion of proof-in-T (and, hence, of such related notions as the consistency of T). If, on the other hand, the latter is the case, then why should one try to capture, as theorems of T, as many truths as is possible concerning the notion of proof-in-T? Evidently, if representation of this notion can occur *without* capturing all such truths, then only certain truths are crucial to the ability of a formula to express it. And if that is the case, then the important thing is to capture *those* truths, and not to capture as many truths as is possible.

It seems, therefore, that Mostowski's proposal is incoherent. If proper arithmetization of a metamathematical notion requires the registration of all of its truths as theorems of T, then, generally speaking, proper arithmetization simply cannot be attained. If, on the other hand, proper representation requires the capturing of only certain truths as theorems of T, then there is evidently no need to capture as many truths "as is possible" as theorems of T. In either event, no cogent defense of the Derivability Conditions can be based on the claim that arithmetizations satisfying those conditions codify more intuitive truths of T's metamathematics than those that do not. For there is no longer any compelling reason to prefer arithmetizations that do this.

In addition to this, there is another failure of cogency that is worth noting. And that concerns the fact that, when judged by Mostowski's standards, arithmetizations satisfying the Derivability Conditions do the worst job possible of representing the notion of unprovability-in-T, and hence the notion of T's consistency. For if an arithmetization satisfies the Derivability Conditions, then it cannot codify any truths regarding the notion of unprovability-in-

T as theorems of T.[15] Thus, judged according to Mostowski's own criterion, the Derivability Conditions would appear to have as many disadvantages as advantages. One would expect to encode more truths concerning the crucially important notion of unprovability-in-T by violating the Derivability Conditions, than by adhering to them. Thus, if the quality of a given representation is determined by the proportion of the relevant intuitive truths that it registers as theorems of T, then arithmetizations satisfying the Derivability Conditions would appear to do a worse job of representing the notion of unprovability-in-T than arithmetizations which violate those conditions. That being the case, one whose ultimate interest is that of finding an apt representation of the notion of T's consistency would be better advised to violate than to satisfy the Derivability Conditions.

As a result of this, it would appear either that Mostowski must give up holding his expressed criterion as a standard for arithmetization generally (i.e., as a standard for the correct representation of such notions as unprovability-in-T as well as for the notions of proof-in-T and provability-in-T), or that he must have some grounds for regarding correct representation of such notions as proof-in-T and provability-in-T as more weighty concerns than correct representation of such notions as unprovability-in-T and the consistency of T, or, finally, that adherence to his standard does not dictate a preference for arithmetizations satisfying the Derivability Conditions.

None of these alternatives is very encouraging to one hoping to obtain a Mostowskian solution to the Stability Problem. The first is tantamount to saying that the Mostowskian proposal is not to be taken as an account of how to represent the notion of unprovability-in-T. And that, in turn, is just to say that Mostowski's account fails to address what, for present purposes, we most want an account of arithmetization to address; namely, the conditions necessary for the proper representation of the notion of T's consistency.

The second alternative is just as bad, since it is based on the mystifying idea that correct representation of the notion of T's consistency is somehow more intimately tied to correct representation of the notions of proof-in-T and provability-in-T than it is to adequate representation of the notion of unprovability-in-T.

Finally, the third alternative must be rejected by the Mostowskian because it amounts to admitting that, judged according to his standards, there simply is no reason to expect that every formula expressing CONSIS(T) will be unprovable in T. Admitting this is the same as admitting that no solution to the Stability Problem has been given, and yet that is just what the Mostowskian proposal is supposedly designed to accomplish.

One might reply to this criticism by calling for a (slight?) modification of Mostowski's criterion. The idea behind this modification is to peg the adequacy of a given T-theoretic representation of a metamathematical notion M not just to the truths regarding M that it causes to be registered as theorems of T, but also to the truths regarding M's constituent notions that its constituent formulae cause to be registered as theorems of T. Thus, for example, the adequacy of '$\sim (\exists x) \text{Prf}_T (x, y)$' as a representation of the notion of unprovability-in-T is to be thought of as depending not only upon the corpus of truths concerning the notion of unprovability-in-T that it encodes as theorems of T, but also upon the corpus of truths concerning the notion of proof-in-T that '$\text{Prf}_T (x, y)$' encodes as theorems of T, and the corpus of truths concerning the notion of provability-in-T that '$(\exists x) \text{Prf}_T (x, y)$' registers as theorems of T.

Thought of in this way, a given formula might qualify as the best representation of a given notion M even if it did not register any of the truths regarding that notion as theorems of T. For it might still have component formulae that do a superior job of registering truths regarding the constituent notions of M as theorems of T and, by dint of that superiority, be the formula with the best overall performance regarding M and its constituent notions. Hence, on the standard proposed by the current modification of Mostowski's criterion, it would be the best representation of M.

Applying this modified criterion, then, the seemingly correct response to our earlier criticism of Mostowski is that it is inconclusive. Consistency formulae satisfying the Derivability Conditions may fail to register any truths concerning the notion of unprovability-in-T as theorems of T. But, the facility of their sub-formulae to register truths concerning the constituent notions of provability-in-T and proof-in-T may still make them the optimal representations of the notion of T's consistency. Formulae violating the Derivability Conditions might register more truths concerning the notion of unprovability-in-T as theorems of T. But if the ability of their constituent formulae to register the truths appropriate to them as theorms of T is inferior, then, all things considered, it might still be best for a consistency formula to satisfy the Derivability Conditions. Consequently, the fact that consistency formulae satisfying the Derivability Conditions cannot register the truths regarding the notion of unprovability-in-T as theorems of T, implies neither defect in those formulae nor defect in the Derivability Conditions as constraints governing our choice of consistency formulae.

This, then, is the argument for the modified version of Mostowski's proposal. In responding to it, I should like to make three points. The first is just a reiteration of a criticism raised against the original proposal; namely, that if the ability of a formula $\mathcal{M}$ to represent a metamathematical notion M really is tied to $\mathcal{M}$'s capacity for encoding truths regarding M as theorems of T, then the discovery that no formula can encode the truths concerning the notion of unprovability-in-T calls the representability of that notion (and the closely related notion of T's consistency) into question. Even in its modified version, the ability of $\mathcal{M}$ to express truths regarding M as theorems of T has something to do with its ability to represent M. And that being so, the failure of a consistency formula to encode truths concerning the notion of unprovability-in-T would appear to have some negative effect on its ability to properly express or represent the notion of T's consistency. The problem for the Mostowskian is, therefore, to show that this does not constitute a disabling impairment.

Our second objection to the modified criterion is related to this problem. For seemingly the only way for the Mostowskian to show that inability to express the truths concerning the notion of unprovability-in-T is not a crippling debility for a consistency formula, is to argue generally that the capacity of a formula $\mathcal{M}$ to express truths regarding M as theorems of T is *not* a crucial determinant of its ability to represent M. Without a general argument to this effect, the Mostowskian would be faced with the unappealing task of having to find relevant differences between particular notions that make representation of some, but not others, critically dependent upon the ability of the representing formulae to codify truths as theorems of T. And, so far as we can see, there is simply no reason to believe that this is so.

It thus appears that if the defender of the modified Mostowskian proposal is to show that failure to capture the truths concerning the notion of unprovability-in-T is not a disabling weakness for a consistency formula, then he must avail himself of a *general* argument to the effect that $\mathcal{M}$'s ability to represent M does not require that $\mathcal{M}$ cause the truths regarding M to be registered as theorems of T. However, appealing to such an argument would place even the modified version of Mostowski's proposal in serious jeopardy. For, like the original, the modified version too puts a premium on $\mathcal{M}$'s ability to codify truths regarding M as theorems of T. Were this not the case, it would be hard to see what could possibly motivate regard for the ability of a component formula of $\mathcal{M}$ to register the truths of a constituent notion of M. In other words, then, the idea behind the modified Mostowskian position seems to be that you represent M by choosing a formula $\mathcal{M}$ whose component formulae represent the constituent notions of M. And, presumably, the test of whether a constituent of $\mathcal{M}$ does a good job of representing a constituent of M is the familiar one; namely, whether it registers the truths concerning that constituent notion as theorems of T.

So, at the level of constituent representation, the modified proposal relies on the old criterion (viz., the ability of a formula to codify the truths of the notion it is to represent as theorems of T).

And it follows from this that even the modified Mostowskian proposal could not tolerate a *general* devaluation of the old criterion. Consequently, the Mostowskian would seem to be forced back into the difficult position of having to say that representation of such notions as unprovability-in-T is *not* governed by the old criterion, even though representation of such notions as proof-in-T and provability-in-T is.

The two objections raised thus far are, in our opinion, sufficient to show the untenability of the current modification of Mostowski's proposal. But our case may be strengthened still further by means of a third and final point. This point has to do with whether the unencodability of truths regarding M should be taken as counting against the representability of M only, or also against the representability of its constituent notions.

Mostowski's explicit claim (cf. p. 25 of Mostowski [1966]) is that M's representation is to be effected by encoding the truths "involving" M as theorems of T. And he makes it very clear that he regards the truths "involving" M as including not only the particular truths which fix its extension, but also the general truths in which it appears as a constituent notion. (So, for example, he regards statements of the form "neither A nor $\sim A$ is provable in T" as statements "involving" the notion of proof-in-T.) Indeed, this is his basis for saying that something more than so-called *strong representation* may be required for adequate representation of a given metamathematical notion.

Evidently, then, Mostowski would count truths concerning the notion of unprovability-in-T as truths "involving" the constituent notions of proof-in-T and provability-in-T. And as a result of this, he would presumably say that the uncodifiability in T of truths concerning the former notion means that there are truths concerning the latter notions that are uncodifiable as theorems of T. Hence, those latter notions can be no more fully representable in T than the former notion. And that spells disaster for the current modification of Mostowski's proposal, since now one can no longer hope to compensate for imperfect codification of the truths concerning the notion of unprovability-in-T itself with supposedly

perfect codification of the truths concerning its constituent notions.

Overall, then, it seems that the current modification of Mostowski's proposal is just as unworkable as the original version. But
before drawing any final conclusions about the Mostowskian
model in general, we need to consider at least one other kind of
modification.

According to this new modification, both the original proposal
and our criticism of it suffer from a failure to distinguish two
importantly different conceptions of metamathematical truth;
namely, the classical and the finitary conceptions. Our criticism of
the original proposal was, at bottom, based on the claim that not
all truths concerning the notion of unprovability-in-T are T-
codifiable (i.e., expressible as theorems of T).[16] And that claim is
cogent so long as our conception of metamathematical truth is the
classical one, since there are all manner of *classically* true metamathematical propositions (e.g., those asserting the consistency of
T, those assreting their own unprovability in T, and those asserting
the undecidability in T of certain formulae) that are not
T-codifiable.

But all this changes once we make the transition from a classical
to a finitary view of metamathematical truth. Upon examination,
we find that all our examples of true but non-T-codifiable propositions presuppose the consistency of T. And, generally speaking,
this presupposition cannot be upheld once the switch from a
classical to a finitary conception of metamathematical truth has
been made, since, for the usual cases of T, we cannot simply
assume that there is a finitary proof of T's consistency.

Thus, the modification of Mostowski's proposal that we now
have in mind makes it possible to eliminate the problematic
restriction of the original version. That is, instead of the incoherent condition calling for the T-codification of as many truths
regarding *M as is possible*, the Mostowskian may now (without
fear of counter-example) avail himself of the more natural and

self-consistent condition which makes M's representation a matter of T-codifying *all* its truths; though here, it is the finitary rather than the classical truths concerning M that are at issue.[17]

On this new version of his model, then, all that the Mostowskian requires for adequate T-theoretic representation of M is the T-codifiability of every *finitary* truth regarding M. But though this frees him from the incoherence of his original position, it raises other problems. In particular, it raises a problem concerning the motivation of F-LPC, as we shall now see.

On the present model, representation of a metamathematical notion M is supposed to consist in the formal replication in T of the finitary thought regarding M. Accordingly, the justification of a principle like LPC is presumably that it enforces the T-codification of finitary truths of the form "$\vdash_T A$".[18] And, applying the same sort of reasoning to F-LPC, we are led to say that its justification is that it calls for the T-codification of finitary propositions of the form "if $\vdash_T A$, then $\vdash_T \text{Prov}_T(\ulcorner A \urcorner)$", all of which are to be regarded as true.

But things are not so simple. For, as is well-known, one may pick '$\text{Prov}_T(x)$' in ways that prohibit that T-codifiability of propositions having this latter form. What is more, one may do so while at the same time satisfying LPC.[19] Therefore, the need to T-codify all finitary truths of the from "$\vdash_T A$" (better, "$P \vdash_T A$") does not lead straightway to the need to codify all propositions of the form "if $\vdash_T A$, then $\vdash_T \text{Prov}_T(\ulcorner A \urcorner)$" as theorems of T. And consequently, the above attempt to justify F-LPC is bogus.

What we learn from this is that the justifiability of F-LPC is independent of that for LPC. Or, to put the same point another way, F-LPC has normative status that is not reducible to that of LPC. One can fully discharge his obligation to T-codify all finitary truths of the form "$\vdash_T A$" (or "$P \vdash_T A$") without thereby engendering any obligation to register all propositions of the form "if $\vdash_T A$, then $\vdash_T \text{Prov}_T(\ulcorner A \urcorner)$" as theorems of T. And because this is so, one cannot derive a duty to satisfy F-LPC from a mere duty to satisfy LPC.

Given that this is true, however, the present modification of Mostowski's proposal is in real difficulty. For it can accord normative status to F-LPC only to the extent that it can find truths of the finitary metamathematics of T to whose T-codification the several instances of F-LPC are to correspond. To do this, one naturally looks to the instances of LPC. But since the normative task for whose execution LPC is wanted (viz., the T-codification of all finitary truths of the form "$P \vdash_T A$") can be carried out *without* turning all of the instances of LPC into finitary truths, it follows that there will inevitably be an element of arbitrariness in any attempt to base the normative status of F-LPC on that of LPC.[20]

When we add to this the fact that, aside from the instances of LPC, there apparently is no fund of metamathematical truths upon which to base F-LPC, the lack of warrant for the current version of Mostowski's proposal becomes clear. It may avoid the internal incoherence of the previous versions, but it offers not the slightest basis for requiring F-LPC as a condition on our choice of consistency formulae. And this by itself is enough to insure that the current version of Mostowski's proposal cannot serve as a solution to the Stability Problem.[21]

In sum, then, we conclude that no version of the Mostowskian model for arithmetization yields any reasonable defense of the Derivability Conditions. But even if some such defense had been found, we would still be left without a full solution to the Stability Problem. For, at most, such a defense would yield only a necessary connection between arithmetizations T-codifying the maximum number of metamathematical truths concerning the notion of T's consistency, on the one hand, and arithmetizations satisfying the Derivability Conditions, on the other. It manifestly would *not* demonstrate any connection between the ability of a formula to semantically express CONSIS(T), and that formula's satisfying the Derivability Conditions. To obtain such a connection, one would have to establish some tie between a formula's ability to T-codify truths concerning a given metamathematical notion M, and its ability to semantically express M. But such a tie, if it exists at all, would not appear to be necessary, since the ability of a formula of T to express a given notion would seem to depend more on the semantical power of T's language than on the deductive power of T.[22] Therefore, the precise character of the connection between T-codification (which requires that truths concerning the represented notion be expressed as theorems of T) and semantical expression would remain unclear, even if a connection between optimal T-codifications and satisfaction of the Derivability Conditions were demonstrated. And we trust that this makes clear just how far short of an adequate solution to the Stability Problem the Mostowskian proposals fall.

5. THE KREISEL–TAKEUTI PROPOSAL

We come now to a rather more epistemological defense of the Derivability Conditions. The central idea of this defense is that the need to satisfy those conditions is founded upon the need of the Hilbertian to know that his formalized ideal theories capture informal mathematical practice. Hence, the argument for the Derivability Conditions is not simply, nor even primarily, that Con(T) must satisfy them in order to semantically express CONSIS(T) for an arbitrarily selected theory T. Rather, it is that, without their satisfaction, we should have no reason to regard T as a reasonable formal approximation of any substantial body of informal mathematical practice. Thus, to return to the familiar terminology of "expression", the claim of the current proposal might best be put this way: satisfaction of the Derivability Conditions is necessary if Con(T) is to be taken as expressing the consistency of any formalism that might reasonably be taken as representing more than a trivial portion of ordinary, informal mathematics.

Kreisel explicitly advocates such a position in at least two places. The first occurs in the context of giving a natural or intuitive statement of a generalized version of G2.

The natural formulation runs quite simply as follows. Given F [= a formal system] and formulae A_1, A_2, A_3 of F, one of the following cannot be derived in F:

A$_1$ expresses ... that F is closed under *modus ponens*, and A$_1$ holds.

A$_2$ expresses that F is complete for numerical arithmetic ... , and A$_2$ holds.

A$_3$ expresses that T is consistent, and A$_3$ holds.

... as far as Hilbert's programme was concerned the absence of A$_1$ or A$_2$ constitutes the *same kind of inadequacy* as of A$_3$: we should have no reason to suppose that F codifies mathematical practice. (Kreisel [1971], p. 118; brackets mine)[23]

The second (which is in some ways clearer and more suggestive than the first) occurs in the context of a claim to the effect that, for purposes of evaluating Hilbert's Program, we need no more refined a generalization of G2 than that originally offered in Hilbert—Bernays [1939], because

... the usual conditions on systems [= the Hilbert—Bernays Conditions] ... are necessary if a formalization of mathematical reasoning is to be *adequate* for Hilbert's programme. ... Let us spell out the two adequacy conditions on a system F:

(a) Demonstrable completeness w.r.t. Σ_1^0 formulae is needed to assure us that elementary mathematics (with a constructive existential quantifier) can be reproduced in F at all ...

(b) Demonstrable closure under cut (and in the quantifier free case also under substitution) is also needed because cut is constantly used in mathematics. Realistically speaking, a (meta)mathematical *proof* of such closure is needed and not a case study of mathematical texts because cut — like most logical inferences — is often used without being mentioned; in contrast, for example, to the use of mathematical axioms. (Kreisel and Takeuti [1974], pp. 34—5; square brackets mine)[24, 25]

The view that emerges from these remarks is, as we indicated earlier, one which holds satisfaction of the Derivability Conditions to be requisite for knowledge that the formal system represented by 'Prov$_T(x)$' codifies ordinary, informal mathematical practice.[26] But now we have a clearer idea of the grounds upon which this view is supposed to rest. There is apparently both a substantive and a procedural ground for each condition of adequacy. The substantive ground for F-LPC is supposedly that informal, ideal practice encompasses elementary mathematics in its entirety, and therefore includes all truths expressed by Σ_1^0 formulae.[27] And its procedural ground is that compliance with this substantive *mathematical* constraint is easily established via metamathematical proof because of the explicitness of informal practice regarding its mathematical content. F-LPC is thus analyzed into a substantive (viz., the Σ_1^0-completeness of T) and a procedural (viz., the provability in T of T's Σ_1^0-completeness) component, and each of these is given its own defense; the former by means of an appeal to the mathematical content of informal, ideal practice, and the latter by means of an appeal to what is epistemologically optimal as a means of establishing the mathematical content of a formal system.

Similarly, F-MP is to be analyzed into a substantive and a procedural element. The substantive element is closure under *modus ponens*, and its defense is a simple appeal to the popularity of *modus ponens* as a technique of informal reasoning. On the procedural side, the constraint is, once again, that compliance with the substantive component be established by means of a metamathematical proof that is codifiable in T. But the defense of the procedural constraint here is taken to be different from what it was in the case of the F-LPC. There, the idea was presumably that the ordinary way in which a formal system is given makes it quite easy to establish facts about its mathematical content with mathematical exactitude.[28] Hence, the attractiveness of a mathematical proof of T's Σ_1^0-completeness. Here, however, the idea is that the facts of informal logical practice don't rise to the same level of explicitness that the facts of mathematical content do, and that because of this, only a T-codifiable proof of T's closure under *modus ponens* can give us the assurance we need that T captures the *logical* practice of informal mathematics.[29]

In sum, then, the Kreisel–Takeuti position consists of both a substantive and a procedural claim. The substantive claim is that if T is to be an adequate formal codification of informal mathematical practice, then it must be both Σ_1^0-complete (so that it captures elementary mathematics), and closed under *modus ponens* (so that it is sure to capture the logical technique of informal classical mathematics). And the procedural claim is that the only practical, or at least the epistemologically optimal, way of coming to know that the above substantive conditions are met is via a T-codifiable metamathematical proof of them. In the argument that follows, we shall call both claims into question.

What would it mean to say that a given system T is an adequate codification of informal mathematical practice (or some branch thereof)? The answer to this question clearly depends upon what one's conception of informal mathematical practice is. And, just as clearly, there are differing conceptions of this. According to one, informal mathematical practice is to be thought of as the totality of assertions and justifications that have gained the popular acceptance of the historically given community of mathematicians. Furthermore, this should be taken to include not only those assertions and justifications that have been explicitly received, but also those whose acceptance would be obligatory for some ideally rational and informed agent sharing the explicit commitments of the actual historical community of mathematicians.

Conceived of in this way, the mathematical *practice* of a group is an idealization of its actual, historical commitments. And such an idealization is quite congenial to idealized conditions like closure under cut and (numerical) arithmetical completeness. So, if what Kreisel and Takeuti mean by mathematical practice is the rounding out of actual historical practice to conditions of perfect rationality and information, then perhaps they are right to say that T cannot hope to codify mathematical practice unless it is closed under cut and arithmetically complete.

Such idealized views of mathematical practice are not without their advantages. For, other things being equal, the stronger and more extreme the idealizations embodied in a given formal system are, the surer it is to capture the relevant area of actual, historical practice. Thus, judged from the standpoint of one whose sole concern is the formal codification of all of a given area of historical practice, the sort of idealization represented by such conditions as closure under cut and Σ_1^0-completeness (and so, the Derivability Conditions) is simply epistemologically optimal strategy (i.e., strategy designed to put one in the best possible position to know that his formalism captures the body of informal practice that it is intended to codify).

But should the appropriateness of the Derivability Conditions be judged from such a standpoint? More specifically, should the appropriateness of the Derivability Conditions, *as constraints governing the Hilbertian's choice of formalizations of ideal reasoning*, be judged from such a standpoint? We think not, since though the ability of his formalisms to capture informal mathematical practice is certainly a concern of the Hilbertian's, it is by no means his only, or even his dominant, concern. This is not to say that he is opposed to idealizing actual practice, for he is not. But his idealizations, being subject as they are to his underlying instrumentalist outlook, have different aims and constraints than those idealizations that we have just been considering. And these differences make such idealizations as closure under cut and Σ_1^0-completeness far less attractive to the Hilbertian than to one whose only concern is that his formalisms be powerful enough to codify mathematical practice.

Chief among the factors which draw the Hilbertian away from excessive idealization are his need to obtain a soundness proof for his formalizations of ideal reasoning, and his natural lack of concern for ideal reasoning which is too long or complex to be of any likely benefit to epistemic devices having the physical and cognitive limitations of human beings. Even the most excessive idealization is no problem for the Hilbertian so long as it poses no threat to his ability to obtain a satisfactory proof of real-soundness. However, as soon as it does, the Hilbertian is prepared to retreat to less excessive idealizations that more nearly approximate the class of instrumentally useful ideal methods.

It would seem, therefore, that in order to defend the likes of closure under cut and Σ_1^0-completeness as appropriate conditions to place on the Hilbertian's formalization of ideal reasoning, one would have to show that without them it is impossible to obtain even a decent approximation to the humanly useful methods of ideal proof. But certainly there is no *a priori* reason to believe that this is so, since the class of humanly useful ideal proofs adheres to length and complexity restrictions which both closure under cut and Σ_1^0-completeness violate.[30] Nor is there any convincing *a posteriori* argument to this effect. Hence, it would seem that we are under no obligation to admit the substantive component of the Kreisel–Takeuti proposal. However, we shall not press the point further at this time, since it is really the procedural rather than the substantive component that we most want to focus our attention on.[31]

As we see it, there are really two distinct procedural claims that Kreisel and Takeuti make. One is that only a *(meta)mathematical* proof can be expected to yield knowledge that a given system T captures informal mathematical practice (or some area thereof). And the other is that only a *T-codifiable* proof can be expected to do this. We count these two claims as distinct because there is no apparent reason why every (meta)mathematical proof should be *T*-codifiable. This point will eventually take on considerable importance for our discussion. But before coming to grips with it, it is necessary to get clearer on the two procedural claims themselves.

We are told that *demonstrable* (i.e., *T*-codifiable) Σ_1^0-completeness is needed to assure us that elementary mathematics can be reproduced (i.e., formalized) in T at all. Likewise, we are told that a *T*-codifiable proof of closure under cut is necessary for knowledge that the logical technique of informal practice is captured by T. But the reasoning offered for this latter claim is rather baffling. The authors begin by claiming that a case-study approach to mathematical texts would run the risk of missing much of the logical technique of informal practice (because logical inference is "often used without being mentioned"). And they immediately conclude from this that the only realistic means of obtaining knowledge that T codifies informal logical technique (in particular, the full range of informal applications of *modus ponens*) is a (meta)mathematical proof of the closure of T under the widely applied principles of logical inference. Finally, they move from this intermediate conclusion to the demand for a *T*-codifiable proof of closure under cut. We shall now consider each of these steps of inference more carefully.

Perhaps what is most bewildering about the first step is that a case-study approach and an approach via (meta)mathematical proof are somehow presented as alternatives. But alternatives with respect to what? The case-study method is supposedly a way of studying *informal* mathematics (i.e., mathematical texts), and the method of (meta)mathematical proof a means of studying *formal-ized* mathematics. To put the point another way, the method of case-study, as applied to the question of logical technique, is designed to be a means of determining what the logical technique of informal mathematical practice really is. The method of meta-mathematical proof, on the other hand, is only intended to determine the logical traits of a given formal system. Since, then, they are not designed to produce knowledge of the same thing, it is hard to see in what sense they might constitute alternatives. It would seem more plausible to say that *both* are need to secure knowledge that a given formalism *T* adequately represents the logical technique of a given body of informal thought; the case-study approach being needed to discern what the logical practice of the given body of informal thought really is, and the meta-mathematical approach being needed to determine whether a given formalism reproduces that technique. Accordingly, let us assume that it is some such account that is to sustain the con-clusion that metamathematical proof is necessary for knowledge that *T* captures the logical technique of a given body of informal mathematics.

But even if we accept this, the legitimacy of the second step of inference remains open to doubt. There is a general doubt, borne of the fact that there is no general reason to believe that everything counting as a metamathematical proof about *T*, will also be codifiable in *T*. Whether a given metamathematical proof is codifiable in *T* would appear to depend both upon which methods of proof are used (e.g., whether they refer to general recursive or only primitive recursive facts about *T*, what forms of induction are used, whether set theoretical methods are used, etc.), and upon the nature of T; i.e., upon which methods of proof it codifies.

And, in addition to these general grounds for doubt, there are more specific grounds which we shall now attempt to elaborate. Basically, these more specific grounds arise from a closer analysis of the instrumentalist conception of theory. Such analysis, in our opinion, reveals a clear basis in instrumentalist thought for opposing the T-provability of the likes of closure under cut and Σ_1^0-completeness. And, in the end, this opposition arises from the distinctive features of the instrumentalist's notion of theoretical revision.

Like the realist, the instrumentalist may find himself in the position of needing to revise his theories. But what revision means for the instrumentalist is not at all what it means for the realist. When an inconsistency is discovered to exist in a realist's theory, his task is ultimately to change the substance of some particular axiom(s) or some particular rule(s) of inference. This is a consequence of the fact that, as a realist, his objective is to construct a system of *true* axioms and *truth-preserving* rules of inference (so as to secure the truth of all his theoretical commitments), and that inconsistency in his theory means that either his axioms are not all true, or his rules of inference are not all truth-preserving (in which case he must either find the guilty culprit(s), or lose the episte-mological basis for his theory in its entirety). Thus, the realist cannot be considered to have dealt adequately with an incon-sistency until he has either dropped some specific axiom(s) or rule(s) of inference.

The instrumentalist, on the other hand, may revise a theory without changing the substance of any axiom or rule of inference, but only the range of its applicability. In other words, upon deriving a contradiction from a set of axioms and rules of inference, his response need not be one of changing some specific axiom(s) or rule(s) of inference, but might rather consist simply in ruling out certain combinations of axioms and rules of inference as forming acceptable (i.e., reliable) proofs.[32] This is a consequence of the fact that he is not committed to calling his axioms true and his rules of inference truth-preserving. He demands instead that a system be an efficient and reliable generator of real truths, where this does not imply that the system be constructed from true axioms and truth-preserving rules of inference. Indeed, for the instrumentalist, the categories of truth and truth-preservingness do not apply to the evaluation of an ideal system. The basic idea, therefore, is that for the instrumentalist (unlike for the realist), discovery of an inconsistency does not force him to change the statement of any particular axiom or rule of inference. In one proof, axioms $A_1, \ldots, A_n$ and rules $R_1, \ldots, R_k$ might lead to the efficient generation of a real truth, whereas in another they might lead to a contradiction. So long as the instrumentalist has a way of isolating the former from the latter, there is no need for him to declare both proofs illegitimate (as revision of the axioms and/or rules of inference would force him to do).[33] And this, as we shall see shortly, has far-reaching implications for the instrumentalists (and, hence, the Hilbertian's) assessment of such conditions as *demonstrable* (i.e., T-codifiable) closure under cut and Σ_1^0-completeness. But before developing these implications further, it is necessary to bring out a corollary feature of the instrumentalist's conception of theory.

This corollary feature is that which gives the instrumentalist the ability to rank ideal proofs in terms of their revisability. Such a ranking might reflect the instrumentalist's attitudes regarding the likely reliability (i.e., real-soundness) of the ideal proofs involved (the less likely to be reliable being the more revisable), his attitudes concerning their relative efficiency or utility as instruments (the less efficient being the more revisable), or (more likely) some combination of the two. But whatever their exact character in any particular case, the plain fact is that the instrumentalist has both access to and motivation to use judgements that would induce him to order his ideal proofs in terms of their revisability.

If this is true, then (with a little idealization) it is reasonable to see the instrumentalist as constructing his theories (i.e., systems of ideal proof) in this way: he starts with a set of axioms and rules of inference, which he treats (in the usual fashion) as determining a

class of would-be proofs; he then attaches a revisability ordering
to this class of would-be proofs.[34] Next, he defines a genuine proof
to be a would-be proof whose conclusion does not conflict with
that of any would-be proof which precedes it in the given
revisability ordering. And, finally, he commits himself to accepting
(as proofs of the theory he is constructing) only the so-called
genuine proofs.[35] In a nutshell, then, the instrumentalist constructs
his theory from a base theory T by imposing a revisability ordering
on the proofs of T, and deleting any proof of T that clashes with
any of its predecessors in this ordering. The deletion of proofs is,
moreover, a purely "extrinsic" affair; that is, it is not connected
with or implicative of any changes in the internal elements (i.e., the
axioms or rules of inference) of the system. (The tolerance of
extrinsic revision is, as we argued above, one of the distinctive
traits of instrumentalism).

The reader will probably have noticed the strong parallels
which exist between the model of instrumentalist theory construction just presented, and a certain variant conception of formalization first presented in Rosser [1936]. This variant model of
formalization starts with an ordinarily conceived formal system T
and adds a consistency clause to its test for proofhood. That is, in
order to count as a proof in T_R (= the Rosser variant of T), a
given sequence $\mathbb{P}$ of formulae must pass not only the test of
proofhood for T, but be such that its end formula neither deny nor
be denied by any of the proofs of T preceeding $\mathbb{P}$ in a given
omega-ordering of the proofs of T.[36]

The parallels between the instrumentalist model of theory
construction presented above and the Rosser variant are no
accident; in fact, the former arose as a direct result of an attempt
to make philosophical sense of Rosser's variant conception of
formalization. And we believe that it does just that. In showing
how the instrumentalist might make use of a revisability ordering
in constructing a system of ideal proof, it motivates the use of a
consistency clause like Rosser's in the definition of a formal
system. And in defending the coherence of an "extrinsic" model of
revision for the instrumentalist, it shows how sense can be made of

a conception of ideal proof which, like Rosser's permits the deletion of proofs from a system, without necessitating a corresponding deletion of axioms or rules of inference. In short, then, we believe that the analysis of instrumentalism presented above provides a philosophical model for Rosser's variant conception of formalization.

To complete our argument we must now draw attention to two features of Rosser systems. The first is that they don't satisfy the Derivability Conditions; specifically, they violate the condition of *demonstrable* Σ_1^0-completeness. Therefore, if Rosser formalization corresponds to a reasonable procedure for the Hilbertian instrumentalist to use in the construction of his ideal theories, then it follows that he is not committed to satisfying the Derivability Conditions. And, specifically, it follows that he is not committed to satisfying the procedural demand of the Kreisel—Takeuti proposal, since it is precisely that which is violated by Rosser formalization.[37] Since, therefore, we take Rosser formalization to represent a potentially reasonable procedure by which the Hilbertian might construct his ideal theories, we conclude that he is under no clear obligation to satisfy the procedural constraint of the Kreisel—Takeuti proposal, and thus that that constraint is unfounded.

The second feature of Rosser systems that we wish to call attention to is their preservation of the *effectiveness* of the notion of proof. That is, the test for proofhood in a Rosser system is an effective procedure (assuming that the test for proofhood in its base theory is).[38] And thus it preserves a feature of formal proof which has traditionally been associated with the Hilbertian's enterprise of metamathematically vindicating the ideal methods. That Rosser systems have this feature appears to depend upon their willingness to work with purely "extrinsic" revision procedures. For if one moves to the use of "intrinsic" procedures (i.e., ones which enforce coherence in the internal apparatus of a system — namely, its axioms and rules of inference), then this quality is typically lost, since there is generally no effective procedure for determining the consistency of a set of axioms or rules of inference.[39]

Thus, it would appear that it is the Hilbertian's instrumentalist epistemology which, in sanctioning the use of extrinsic revision procedures, is fundamentally responsible for his ability to escape the procedural constraint of Kreisel and Takeuti, while at the same time preserving the *formal* character of his systems.[40]

Overall, then, we find little reason for the Hilbertian to accede to the substantive demand of Kreisel and Takeuti, and a solid epistemological reason for him to oppose their procedural demand. Thus, the content of the Derivability Conditions remains unmotivated, and their demonstrability requirement implausible. And so, we can only conclude that the current proposal fails to solve the Stability Problem.

6. THE CLASSICAL PROPOSAL?

The proposal that we shall be examining in this section is believed by some (cf. Prawitz [1981], p. 258) to have a history tracing back to Hilbert's classical writings on his program. In both his 1925 Münster (cf. p. 383) address and his 1927 Hamburg address, Hilbert made remarks concerning consistency proofs and proofs of real-soundness in which he drew a very tight connection between the two.

> To be sure, one condition, a single but indispensable one, is always attached to the use of the method of ideal elements, and that is the proof of consistency; for, extension by the addition of ideal elements is legitimate only if no contradiction is thereby brought about in the old, narrower domain, that is, if the relations that result for the old objects whenever the ideal objects are eliminated are valid in the old domain. . . this problem of consistency is perfectly amenable to treatment. For the point is to show that, when ideal objects are introduced, it is impossible for us to obtain two logically contradictory propositions, $\mathcal{U}$ and $\sim \mathcal{U}$. (Hilbert [1927], p. 471)

Thus, while in one breath (viz., when he says that the task is to show that "the relations that result for the old objects whenever the ideal objects are eliminated are valid in the old domain"), Hilbert's chief concern is clearly with the real-soundness problem, when it comes to giving a precise formulation of his foundational project (cf. the last sentence of the passage just quoted), what he describes is the consistency problem. Such facile interchanging of the two projects naturally encourages one to conclude that Hilbert saw them as basically equivalent. And this, in turn, suggests a new defense for the Derivability Conditions (or, better, of F-LPC or its generalization to provable Σ_1^0-completeness).[41] That defense proceeds as follows.

(Claim 1) Part of Hilbert's original program was to obtain the proof of the real-soundness of a given system T of ideal mathematics from a proof of its consistency.

(Claim 2) In order to obtain a finitary proof of the real-soundness of system T from a finitary proof of its consistency, one must possess a finitary proof of its Σ_1^0-completeness.

From Claim 2 and the presumption that finitary reasoning can be theorem-wise codified in T, it follows that

(Claim 3) in order to obtain a finitary proof of the real-soundness of T from a finitary proof of its consistency, the Σ_1^0-completeness of T must be T-codifiable (and, hence, 'Prov$_T(x)$' must satisfy F-LPC).

It then follows from Claims 1 and 3 that

(Claim 4) if Hilbert's original program is to be carried out, then T must be demonstrably Σ_1^0-complete (and, thus, 'Prov$_T(x)$' must satisfy F-LPC).

Since Claim 4 is based on Claims 1 and 3, and since Claim 3 is based on Claim 2, and, finally, since the grounds for Claim 1 have already been presented, the only thing required for the completion of the above argument is a defense of Claim 2. And so far as I have been able to determine, there are two different, though related, arguments that have been given for this claim.

The first of these[42] identifies real-soundness with the following reflection principle for T.

(Π_1^0-reflection): For any Π_1^0 formula A of T, $\vdash_T \text{Prov}_T(\ulcorner A \urcorner) \supset A$.

It then proceeds to show (roughly) that, with the help of demonstrable Σ_1^0-completeness, it can be proven that for any Π_1^0
formula A of T,

$$(*) \qquad \mathrm{Con}\,(T) \vdash_{\mathrm{PRA}} \mathrm{Prov}_T(\ulcorner A \urcorner) \supset A.^{43}$$

Presumably, one is supposed to conclude from this first that
without demonstrable Σ_1^0-completeness, one could not prove (*),
and thence that, without finitary proof of T's Σ_1^0-completeness, it
would be impossible to obtain a finitary proof of T's Π_1^0-sound-
ness (and thus its real-soundness, generally) from a finitary proof
of its consistency.

The second argument, which is suggested in various places by
Prawitz (cf., Prawitz [1971], pp. 238—9; [1972], p. 128; [1981],
pp. 257—8),[44] is carried out at the level of intuitive, rather than
arithmetized, metamathematics. He says that the real-soundness of
T can be inferred from its consistency, given two auxiliary
premises: namely, (i) that every true real formula of T is provable
in T, and (ii) that the denial of a false real sentence is a true real
sentence. With these two premises in hand, the argument presumably proceeds as follows: if T is not real-sound, then there is
some real formula R of T such that R is provable in T, but false;
but then, assuming (ii), the denial of R will be true, and so, by (i),
provable in T; hence, T will be inconsistent. Thus, if T is consistent and real-complete, and if the denial of a false real formula
is a true real formula, then T is real-sound. Such, at any rate, is the
argument that Prawitz seems to favor.

But where is Σ_1^0-completeness supposed to enter the Prawitzian
argument? Presumably, through the joint action of (i) and (ii). For,
assuming that A is a true Σ_1^0 formula, it follows that $\sim A$ is
(equivalent to) a false Π_1^0 formula. Thus, by (ii), and the fact that
$\sim A$ (being Π_1^0) is a real formula, we may conclude that $\sim \sim A$ is a
true real formula. And so, by (i), it follows that $\vdash_T \sim \sim A$, and
therefore that $\vdash_T A$. Hence, it follows from (i) and (ii) that T is
Σ_1^0-complete.[45]

There are various points that one might make in response to these
arguments. For example, in connection with the first argument, we

might mention the dubious identification of 'Prov$_T$($\ulcorner A \urcorner$) $\supset$ A' (for Π^0_1 cases of A) with real-soundness. How is it that this formula is supposed to express the claim that every real Π^0_1 formula provable in T is finitarily true (i.e., finitarily provable)? In the first place, it is only a schema, and so does not express a general proposition at all. And secondly, even its instances do not express the *finitary* truth of the various Π^0_1 theorems of T, but rather only their *classical* truth. Therefore, the first argument would not appear to show that a finitary proof of T's Π^0_1-soundness can be obtained from finitary proof of its consistency, given a finitary proof of its Σ^0_1-completeness. And, consequently, it doesn't define any clear and cogent role for demonstrable Σ^0_1-completeness to play in the allegedly Hilbertian project of deriving T's real-soundness from its consistency.[46]

As regards the second argument, perhaps our most serious reservation concerns the use of assumption (ii); i.e., the claim that the denial of a false real sentence is a true real sentence. It is acceptable when restricted to variable-free real sentences. But it would seem to be false when applied to Π^0_1 sentences; at least if it is correct to assume that some Π^0_1 sentences (e.g., '(x) $(x = 0)$', or '(x) $(x = x')$') are to be counted as false. For the denials of such sentences are (equivalent to) Σ^0_1 sentences and, as is well-known, Hilbert didn't countenance such sentences as real sentences at all, much less as true ones. So, assumption (ii) seems quite suspect as a doctrine supposedly applying to the evaluation of Hilbert's original program.

Yet even if we accept assumption (ii), the argument has its faults. For it seems only proper to conceive of real-soundness and real-completeness in terms of finitary provability rather than classical truth. (Understood in this way, the real-soundness of T demands not the classical truth, but rather the finitary provability of every real sentence provable in T. Likewise, T's real-complete ness calls for the provability in T of every finitarily provable real sentence, and not of every classically true real sentence.) But if soundness and completeness are conceived of in this way, Prawitz' argument fails, since we can no longer make our way from the hypothesis that T is not real-sound, to the claim that there is a false real sentence provable in T. Rather, all we can say is that there is a real sentence R, provable in T, but not finitarily provable. And since no one (not even the pre-Gödelian finitist who would have classified such an inference as invalid for anything other than variable-free real sentences), would want to infer from this that the denial of R *is* finitarily provable, it follows that real-completeness cannot be invoked to conclude that the denial of R is provable in T. It would seem, therefore, that what the Prawitzian argument needs is not (i) and (ii), but rather a principle to the effect that the denial of any finitarily unprovable real sentence is a theorem of T. But, we see no way of motivating such a requirement.

All in all, then, the case for Claim 2 strikes us as dubious. But we don't want to make too much of this, since doing so might distract the reader from our main argument, and since we are not sure that we have correctly understood Prawitz. So, let us suppose that a satisfactory response to all of the points raised above can be given. Indeed, let us simply grant the truth of Claim 2 — and that of Claim 4 as well! We can still make our main case. For the charge which we wish to bring against the current proposal is not one of either falsehood or unfoundedness, but rather one of impotence.

On the current proposal, the role assigned to the Derivability Conditions (specifically to demonstrable Σ_1^0-completeness, or F-LPC) is that of securing a link between proofs of T's consistency and proofs of its real-soundness: specifically, it is that of guaranteeing the derivability of a finitary proof of T's real-soundness from a finitary proof of its consistency. The cost incurred by giving up the likes of F-LPC or demonstrable Σ_1^0-completeness is, therefore, precisely that of relinquishing such a guarantee. And while we would not want to say that such a cost is entirely trivial or insignificant, we would nonetheless insist that it is much less than that which is typically attributed to G2. For, on the current defense of F-LPC, the generalized version of G2 cannot be taken as showing that either T's consistency or its real-soundness is finitarily unprovable.[47] Rather, it shows at most that a finitary proof of the latter cannot be derived from a finitary proof of the former. And this is surely a much less damaging consequence for Hilbert's Program than that which is usually ascribed to G2.

The current defense of the Derivability Conditions thus yields G2 in a form bearing only relatively minor consequences for the Hilbertian instrumentalist. It may oblige him to give up any hope of turning a proof of T's consistency into a proof of its soundness, but it does not require that he abandon the search for a finitary proof of T's soundness. Consequently, it does not defeat his proposed strategy for dealing with the Dilution Problem. And this, to us, is a fact of considerable importance.

7. CONCLUSION

We have examined three proposed solutions to the Stability Problem, and found them all to be insufficient in one respect or another. It is possible, of course, that there be other more convincing ways in which a link between a formula's ability to express CONSIS(T) and its satisfaction of the Derivability Conditions might be forged. But, frankly, we are skeptical. For it seems to us that the phenomenon of Rosser formalization, with its clear instrumentalist motivation, calls the very existence of such a link into question.[48] Thus, the argument of this chapter has, in our opinion, done more than merely defeat the extant defenses of the Derivability Conditions. It has, in addition, exposed a natural gap between the Hilbertian's instrumentalist epistemology and satisfaction of the Derivability Conditions. Because of this, there would seem to be little chance of a solution to the Stability Problem.

NOTES

[1] The proof given here generally follows that of Smorynski [1977].
[2] '$\ulcorner G \urcorner$' stands for the standard numeral of T which, under the usual interpretation of T's language, denotes the Gödel number of the formula G.
[3] LPC and LPS are generally not treated as separate conditions. Rather they are combined into a single condition which is most often referred to as the *weak representability* of T's theorems (rather, theorem-numbers) by 'Prov$_T(x)$'. Our

reason for splitting this condition up is that only LPC is needed for our proof of G2. Generally speaking, a set S of n-tuples of numbers is *weakly representable* in T just in case there is a formula '$F(x_1, \ldots, x_n)$' of T such that for every n-tuple of numbers $\langle k_1, \ldots, k_n \rangle, \langle k_1, \ldots, k_n \rangle \in S$ iff $\vdash_T F(\bar{k}_1, \ldots, \bar{k}_n)$.

[4] 'Con(T)' stands for a formula of the form '$\sim \text{Prov}_T(\ulcorner \oplus \urcorner)$', where '$\ulcorner \oplus \urcorner$' is a term of T recognized as denoting (the Gödel number of) some absurdity.

[5] The idea for a proof of a generalized version of G2 using only the conditions we use is traceable back to the work of Löb [1955]. In that paper Löb offers a (positive) solution to a problem posed by Henkin (namely, whether a certain formula asserting its own provability is a theorem of the standard formalisms for recursive arithmetic). To do so, he proved the following more general result which has subsequently come to be known as Löb's Theorem.

> For any formula A of the language of T, $\vdash_T \text{Prov}_T(\ulcorner A \urcorner) \supset A$ only if $\vdash_T A$.

To prove this theorem he began with the following set of Derivability Conditions originally formulated in Hilbert–Bernays [1939].

I.   For any formulae A, B of T,
$\vdash_T \text{Prov}_T(\ulcorner A \supset B \urcorner) \supset [\text{Prov}_T(\ulcorner A \urcorner) \supset \text{Prov}_T(\ulcorner B \urcorner)]$.

II.   For any formulae A, B of T,
if $A \vdash_T B$, then $\vdash_T \text{Prov}_T(\ulcorner A \urcorner) \supset \text{Prov}_T(\ulcorner B \urcorner)$.

III.   If '$f(x)$' is a recursive term of T, then
$\vdash_T f(x) = 0 \supset \text{Prov}_T(\ulcorner f(x) = 0 \urcorner)$.

IV.   For any formula A of T, if $\vdash_T A$, then $\vdash_T \text{Prov}_T(\ulcorner A \urcorner)$.

From II and III he then derived the following consequent condition.

V.   For any formula A of T, $\vdash_T \text{Prov}_T(\ulcorner A \urcorner) \supset \text{Prov}_T(\ulcorner \text{Prov}_T(\ulcorner A \urcorner) \urcorner)$

And using I, IV, and V (which are just our F-MP, LPC and F-LPC) plus another form of diagonalization for sentences asserting their own provability, he proved his theorem.

A generalized form of G2 is, of course, readily obtained from Löb's Theorem. For if we let $\oplus$ be some absurdity (i.e., some formula such that if T is consistent, then not-$\vdash_T \oplus$, then Löb's Theorem allows us to infer that

(*)   not-$\vdash_T \text{Prov}_T(\ulcorner \oplus \urcorner) \supset \oplus$.

And from (*) and the assumption that T's logic is classical it follows that

(**)   not-$\vdash_T \sim \text{Prov}_T(\ulcorner \oplus \urcorner)$.

It appears, therefore, that the historical origins of the current generalization of G2 are attributable in large measure to Löb. As we shall see in the next note, however, the Derivability Conditions necessary for a proof of Löb's Theorem (viz., a set including F-MP) may be stronger than those necessary for a generalized version of G2. (This point is made in Kreisel and Takeuti [1974], pp. 7–8 in response to the work of Jeroslow described in the next note).

[6] Strictly speaking, it isn't true to say that all of the present conditions are needed in order to obtain a generalized version of G2. For in Jeroslow [1973], it is shown how to prove such a result without appealing to F-MP. His proof is based on a new self-referential formula J, which asserts the provability of its negation, and a strengthened form of diagonalization by which the *term* denoting J is provably *identical to* the *term* denoting 'Prov$_T$($\ulcorner{\sim}J\urcorner$)'. This obviates the need for an appeal to F-MP because *modus ponens* (and hence, F-MP) is only required by the need to infer the provability of the right side of the diagonalization lemma (viz., '${\sim}$Prov$_T$($\ulcorner G\urcorner$)') from that of the left (viz., G). But since, in Jeroslow's proof, diagonalization is not a provable *equivalence* between *sentences*, but rather a provable *identity* between *terms* (that designate sentences), the need for a principle of *sentential* inference like *modus ponens* never arises.

However, though Jeroslow's generalization of G2 can do without F-MP, it cannot do without LPC and F-LPC. Thus, since our main concern is with these latter two conditions (most particularly with F-LPC) most of what we have to say about generalizations of G2 based on the current set of Derivability Conditions will apply as well to Jeroslow's generalization. I would like to thank Robert Jeroslow, Michael Byrd, and George Boolos for useful discussions of Jeroslow's work.

[7] Of course, in the SA, P1 of the present inference does not appear as a premise itself, but rather as the consequent of a premise whose antecedent is the proposition that T is consistent. And, as a result, the conclusion of the current inference is also not drawn in the form shown, but rather appears as the consequent clause of a conditional proposition whose antecedent clause is the proposition that T is consistent. However, this doesn't change the fact that the advocate of the SA is committed to the validity of the present inference, and it is this that is the matter of chief importance for the moment.

[8] We should also mention that we won't often bother to be explicit about the distinction between the finitary proposition that T is consistent and the classical proposition that T is consistent. Most often what we have to say will either apply to both equally, or the context will make clear which is the intended sense.

[9] That LPC does not entail (much less presuppose) that 'Prov$_T$($\ulcorner A\urcorner$)' semantically expresses the proposition that A is provable in T is easy to show. Consider, for example, the formula '$(\exists y)$ ($\ulcorner A\urcorner = y \lor \ulcorner A\urcorner \neq y$)'. It satisfies LPC, but clearly cannot be considered to semantically express the proposition that A is provable in T, since (among other things) it produces a consistency formula whose *denial* is *logically* provable!

[10] The strict parallel to LPC is ruled out by the following proof. Suppose that there were a strict parallel to LPC (i.e., suppose that for all A, if not-$\vdash_T A$, then $\vdash_T {\sim}$Prov$_T$($\ulcorner A\urcorner$)). Then, assuming that T is consistent, and hence that not-$\vdash_T G$ (where G is the undecidable formula of G1), it follows that $\vdash_T {\sim}$Prov$_T$($\ulcorner G\urcorner$). And so, by DIAG, it would also follow that $\vdash_T G$; which, of course, contradicts our earlier assumption that T is consistent. Thus, if T is consistent, then the strict parallel to LPC cannot be true.

¹¹ In order to call the reliability of a statistical or inductive correlation between formulae expressing CONSIS(T) and formulae satisfying the Derivability Conditions into question, we do not believe that it is necessary (nor even necessarily helpful) to produce an actual example of a formula expressing CONSIS(T) that does not satisfy the Derivability Conditions. One can (and sometimes can *only*) attack inductively based generalizations in other ways; e.g., by pointing out our inability to give them a meaningful place in any larger explanatory scheme. And, in general, this is the sort of approach that we shall be taking to the issue of the statistical connection between formulae satisfying the Derivability Conditions and formulae expressing provability. However, in the section on the Kreisel-Takeuti Proposal, we shall offer a description of a class of provability formulae that actually violate the Derivability Conditions.

¹² Mostowski is here speaking explicitly about the T-theoretic representation of an informal mathematical notion. But his remarks can clearly be extended to apply to the T-theoretic representation of informal metamathematical notions as well. In order to make this extension, all that is needed is the presupposition of some Gödel numbering whereby statements of the metamathematics of T are semantically linked with the informal mathematical statements that the formulae of T directly express. Mostowski is suggesting, then, that we choose a Gödel numbering which "maps" intuitive truths (falsehoods) of T's metamathematics to truths (falsehoods) of informal number theory, and that we adopt a formalization scheme which, insofar as is possible, registers these informal number-theoretic truths as theorems. In that way, maximizing the body of truths concerning X (X being the informal number-theoretic notion which, via Gödel numbering, encodes the metamathematical notion M) that are expressed as theorems of T, is equivalent to maximizing the body of truths concerning M that are registered as theorems of T. Hence, Mostowski's statement, when properly interpreted, is really a statement about the proper representation of metamathematical, as well as mathematical notions.

¹³ A set S of n-tuples of numbers is said to be *strongly represented* by the n-ary formula '$F(x_1, \ldots, x_n)$' of T when, for every n-tuple of numbers $\langle k_1, \ldots, k_n \rangle$, $\langle k_1, \ldots, k_n \rangle \in S$ only if $\vdash_T F(\bar{k}_1, \ldots, \bar{k}_n)$ and $\langle k_1, \ldots, k_n \rangle \notin S$ only if $\vdash_T \sim F(\bar{k}_1, \ldots, \bar{k}_n)$.

¹⁴ The illustration which Mostowski cites in support of this is taken from Rosser [1936], and consists in the fact that not all truths of the form '$\sim \text{Prov}_T(\ulcorner A \urcorner)' \wedge \sim \text{Prov}_T(\ulcorner \sim A \urcorner)$' are provable in T. But one could just as well take G or Con(T) as examples.

¹⁵ Actually, this isn't quite accurate since, in addition to satisfaction of the Derivability Conditions, we must also assume that

(*)   for any A, $\vdash_T \text{Prov}_T(\ulcorner \oplus \urcorner) \supset \text{Prov}_T(\ulcorner A \urcorner)$.

However, with (*) at our disposal, it is possible to show that if '$\text{Prov}_T(x)$' also satisfies the Derivability Conditions, then there is no A such that not-$\vdash_T A$ and $\vdash_T \sim \text{Prov}_T(\ulcorner A \urcorner)$. For if we suppose that there is an A such that both

(1)   not-$\vdash_T A$

and

$$(2) \vdash_T \sim \text{Prov}_T(\ulcorner A \urcorner),$$

then it follows (by (2), (*) and F-MP) that

$$(3) \qquad \vdash_T \sim \text{Prov}_T(\ulcorner \oplus \urcorner).$$

And from the assumption that '$\text{Prov}_T(x)$' satisfies the Derivability Conditions, it follows (by Gen G2) that

(4)   if T is consistent, then not-$\vdash_T \sim \text{Prov}_T(\ulcorner \oplus \urcorner)$.

But (3) and (4) together imply that T is inconsistent, and this contradicts (1). It thus follows that no true statement of unprovability can be codified as a theorem of T using any provability formula satisfying (*) and the Derivability Conditions.

Actually, the addition of (*) is quite minimal, since it follows from the Derivability Conditions (in particular, LPC and F-MP) given ordinary assumptions about the language and logic of T. Under such assumptions, we know both that

(i)   for any formula A of T, $\vdash_T (\oplus \wedge \sim \oplus) \supset A$,

and that

(ii)   for any formulae A, B, C of T,
        if $\vdash_T (A \wedge B) \supset C$ and $\vdash_T B$, then $\vdash_T A \supset C$.

Furthermore, given the characterization of $\oplus$ (viz., that it is an anti-theorem of T), it must be the case that

(iii) $\vdash_T \sim \oplus$.

But from (i)–(iii) it follows that

(iv)   for any formula A of T, $\vdash_T \oplus \supset A$.

And from (iv) and LPC, it follows that

(v)   for any formula A of T, $\vdash_T \text{Prov}_T(\ulcorner \oplus \supset A \urcorner)$.

Therefore, since by F-MP we also know that

(vi)   for any formula A of T,
        $\vdash_T \text{Prov}_T(\ulcorner \oplus \supset A \urcorner) \supset (\text{Prov}_T(\ulcorner \oplus \urcorner) \supset \text{Prov}_T(\ulcorner A \urcorner))$,

it immediately follows (from (v) and (vi)) that (*) is true.

Thus, though strictly speaking (*) doesn't follow from the Derivability Conditions, it does follow given very ordinary and plausible assumptions about the language and logic of T.

[16] To recapitulate briefly, the most natural and self-consistent version of the Mostowskian standard of representation would demand that *every* truth regard-

ing M be T-codified. The discovery that this standard cannot be met for certain basic metamathematical notions is what drives the Mostowskian to the ultimately incoherent standard of requiring the T-codification of only as many truths *as is possible*.

[17] That this is a modification of Mostowski's own position, is clear from the examples he cites of metamathematical truths that are not codifiable as theorems of T. Citing Rosser [1936], Mostowski notes that the set $\mathscr{U}$ of (Gödel numbers of) formulae that are undecidable in T is not recursively enumerable. From this it follows that $\mathscr{U}$ is not weakly representable in T, and so we know that not all intuitive truths of the form "neither A nor $\sim A$ is provable in T" can be expressed as theorems of T.

But since the truth of such propositions depends upon the consistency of T, and since Mostowski makes no claim for the finitary provability of T's consistency, it seems clear that he is putting them forth as truths of the *classical* rather than the *finitary* metamathematics of T.

[18] Actually, a little refinement is needed here, since statements of the form "$\vdash_T A$" are existential in character and, therefore, would not have been counted by Hilbert as genuinely finitary propositions at all. However, LPC follows (classically) from the codification in T of genuinely finitary truths of the form "$P \vdash_T A$" (where 'P' may be taken as standing for a finite sequence of formulae of T), and so may be taken as motivated by the need to codify *those* truths as theorems of T.

[19] An example of this is the symmetrical version of the Rosser provability predicate studied in Kreisel and Takeuti [1974] (cf. pp. 15–16, 46–8).

[20] One might attempt to derive F-LPC from some stronger condition like provable Σ_1^0-completeness which, in turn, would be founded upon a strengthening of LPC. Ultimately, however, this is no more satisfactory than the current proposal, since one can codify existential truths of T's metamathematics without turning the corresponding conditionals (linking the existential truths to their T-codifications) into finitary truths.

We shall have considerably more to say about strengthenings of F-LPC in subsequent sections of this chapter.

[21] Could one simply add to the demand that all finitary truths of the form "$P \vdash_T A$" be T-codified, the further (evidensory) demand that the T-codifiability of each such truth be itself finitarily provable, and thus obtain a defense of F-LPC? Not without arbitrariness. And perhaps not without some incoherence as well. This comes from the fact that using the Derivability Conditions, one can obtain the following formalized version of G2.

$$\vdash_T \sim \mathrm{Prov}_T(\ulcorner\oplus\urcorner) \supset \sim \mathrm{Prov}_T(\ulcorner\sim\mathrm{Prov}_T(\ulcorner\oplus\urcorner)\urcorner).$$

And, applying the above-mentioned evidensory standard consistently (i.e., applying it not only to the T-codifiability of truths of the form "$P \vdash_T A$", but also to the T-codifiability of truths of the form "not-$\vdash_T A$"), one would be led to a principle to the following effect.

$$\vdash_T \sim \mathrm{Prov}_T(\ulcorner\oplus\urcorner) \supset \mathrm{Prov}_T(\ulcorner\sim\mathrm{Prov}_T(\ulcorner\oplus\urcorner)\urcorner).$$

But from this principle and the above formalization of G2, it follows that

$$\vdash_T \sim \text{Con}(T).$$

And from this, in turn, it would appear to follow (at least for the usual cases of T) that $\text{Con}(T)$ does not express $\text{CONSIS}(T)$. Hence, the Derivability Conditions (which produce the formalization of G2) together with the general evidensory requirement that the T-codifiability of a notion by a formula be finitarily provable do not seem to be compatible as conditions to be placed upon the formula to be used as an expression of $\text{CONSIS}(T)$.

[22] Bezboruah and Shepherdson [1976] attribute to Kreisel a view of expressibility which would link it primarily to deductive rather than expressive power.

Kreisel has argued . . . that something like the Hilbert–Bernays conditions represent a completely satisfactory statement of Gödel's second theorem on the grounds that $\text{Th}(x)$ can hardly be considered to express the notion 'x is the g.n. of a theorem of L' if such simple properties of proofs are not formally provable. (p. 503)

And they go on to voice their argreement with this by saying that in (deductively speaking) weak systems like Q, the consistency formula does not express consistency but only an "algebraic property", and the provability formula really does not express any syntactical property at all.

We are not sure what notion of "expression" these authors may be considering, but it seems to us a mistake to run the question of whether a given formula of T semantically expresses (*modulo* a Gödel numbering and the standard interpretation of T's language) a certain metamathematical proposition together with the quite different question of T's deductive strength. Of course, to finally make an inference from the unprovability in T of $\text{Con}(T)$ to the finitary unprovability of $\text{CONSIS}(T)$, one needs both types of information (i.e., one needs to know both that $\text{Con}(T)$ semantically expresses $\text{CONSIS}(T)$, and that T is deductively powerful enough to theorem-wise subsume finitary reasoning). But in an investigation of the status and ultimate justification of the Derivability Conditions, failure to distinguish these two different types of questions only leads to confusion.

[23] We are perplexed by Kreisel's juxtaposition of the consistency requirement with the requirements of closure under *modus ponens* and arithmetical completeness. Why should failure to produce a T-codifiable proof of T's consistency be taken as in any way signifying the inability of T to codify mathematical practice? That T be an adequate codification of informal mathematical practice, and yet not have a T-codifiable consistency proof would seem to be well within the range of conceivability. And this applies to the Hilbertian as well as to the non-Hilbertian.

[24] Condition (a) (i.e., demonstrable Σ_1^0-completeness) is, of course, just a generalization of F-LPC, since '$\text{Prov}_T(x)$' is a Σ_1^0 formula. Likewise, condition (b) is just our F-MP.

[25] I am unclear as to why Kreisel should go out of his way to include the qualifying clauses "with a constructive existential quantifier" and "in the

quantifier free case also under substitution" in the statements of conditions (a) and (b), respectively. These qualifications make rather more sense for one who is seeking an adequate codification of finitary or constructive reasoning than for one (like the Hilbertian) whose immediate concern is for a codification of informal *ideal* reasoning.

It may also be worth noting that the *T*-provability of *T*'s consistency, whose presence alongside (a) and (b) in Kreisel's previous list of adequacy conditions so perplexed us, is now absent.

26 Perhaps this is as good a time as any to take stock of Kreisel's views concerning the nature of a formal system. In general, his view is that different tasks dictate different conceptions. For some (e.g., model-theoretic tasks generally) it may be sufficient to identify a formal system with its set of theorems. But for others, including the task of assessing Hilbert's Program, a more discriminating conception is said to be necessary. With respect to these latter tasks, Kreisel remarks that

... it is not sufficient to identify a formal system with its set of deductions [much less its set of theorems], we have to consider the manner in which we verify that a syntactic object is a deduction, in short we have to consider the formal rules. (I shall take these rules to be given by their production rules.) (Kreisel [1971], p. 117; square brackets mine)

And again

Which data determine a formal system, say *F*? ... it is evident that, for consistency questions, the proper data are determined by the *rules* of *F* and not, for example, by the set of theorems generated by *F*. (Kreisel [1976], p. 113).

For related statements, see Kreisel [1958a], pp. 289–90; Kreisel [1968], p. 363; Kreisel and Levy [1968], pp. 123–4; and Kreisel [1965], pp. 153–4.

27 Obviously, the defense of the substantive component of F-LPC is nothing other than a defense of LPC (or, better, its generalization to Σ_1^0-completeness).

28 I am speculating here. Kreisel and Takeuti offer no explicit defense of the procedural component of (a). But what I am saying seems to be in keeping with the spirit of their position.

29 It is also worth noting that this lack of explicitness in the logical practice of informal mathematics may form part of the basis for the substantive component of condition (b). Since the use of *modus ponens* is supposedly so widespread, and since so much of that use may be tacit, it may be that the only practical (or at least the best) way we have of insuring that *T* captures the logical technique of informal mathematics is to *close* it under *modus ponens*.

30 In systems where cut is a rule of inference, it will eventually happen that the combined length/complexity of the proof of '*A*' and the proof of '*A* ⊃ *B*' will exceed that which is humanly useful. And for systems closed under cut but not containing it as a rule of inference, it is generally possible to show that there are theorems whose proof is unfeasibly lengthened just because cut is absent. Likewise, with respect to Σ_1^0-completeness, it needs to be said that Σ_1^0 truths can grow so long and complex that their shortest and simplest proofs in the given system are of unfeasible length/complexity.

[31] Nor are we sure how much further the point *might* be pressed, since (as was noted in Chapter III) such conditions as those in question might be required in order to afford a feasible inductive approach to the soundness problem for the system in question. But even if this is so, it suggests a radically different defense of the Derivability Conditions than the one being proposed by Kreisel and Takeuti. And so it does not support their claims.

[32] The simplest and most direct case of such exclusion would be that which rules out the particular anomaly-producing "proof" *itself* (or, where the anomaly is produced not by one, but rather by a pair of proofs, that which rules out one of them).

[33] It is a fact that every inconsistent system has a "least" inconsistency (i.e., an inconsistency whose proof is at least as short as any proof of an inconsistency in the system). Now suppose that in constructing the proofs of an inconsistent system *I*, one can generate every instrumentally useful proof of the system without ever constructing a proof as long as that of its least inconsistency. The instrumentalist advocate of *I* could, in such a case, simply "rule out" all proofs long enough to be a proof of an inconsistency, without (necessarily) ruling out any axioms or rules of inference of the system. Indeed, it might be that for every axiom and rule of inference of *I*, there be some useful proof in which it appears, so that the instrumentalist advocate of *I* would have a motive for *not* dropping any of them from the system. Still, he could, at least in principle, cope with inconsistency by ruling out *combinations* of the axioms and/or rules without which the proof of the "least" inconsistency could not be generated. And he might be able to do this without eliminating any combination having epistemic utility as a generator of real truths (e.g., he might be able to get by just by declaring any combination as lengthy as the shortest proof of the least inconsistency illegitimate). But more on this later.

[34] For present purposes, we may assume that this ordering reflects only the "ordinal" characteristics of the agent's judgements of relative revisability. But we must assume also (and this is one point where some idealization is necessary) that some total ordering of the would-be proofs is an acceptable representation of the agent's (idealized) judgements of relative revisability. Finally, we also need to idealize to the point where we can say that the revisability ordering is such as to give only finitely many "predecessors" (all of which are recognized) to each of its elements.

While we would not go so far as to say that these idealizations are somehow natural or obvious, we would at least maintain their coherence; i.e., we know of nothing that would rule them out as unreasonable. What we are saying, then, is that the current assumptions constitute a reasonable (though idealized) conception of a revisability ordering.

[35] In the end, we don't think that the instrumentalist would commit himself to something even this strong, since not every "genuine" proof (in the current sense of the term) is going to be a gainful epistemic instrument. Thus, the class of genuine proofs is probably best seen as forming an "upper bound" on the class of ideal proofs to whose defense the instrumentalist would, strictly speaking, be committed.

[36] Our definition of a Rosser variant is not exactly the same as that which was originally introduced in Rosser [1936]. That definition was as follows: D is a proof in T_R just in case D is a proof in T and, for every proof D^* that precedes D in the given omega-ordering of the proofs of T, the conclusion of D^* is not the negation of the conclusion of D.

Our definition places a "symmetry" requirement on the relation between the conclusions of D and D^* that is absent in Rosser [1936]. That is, we require *both* that the endformula of D not be the negation of the endformula of D^* *and* that the endformula of D^* not be the negation of the endformula of D; whereas, Rosser requires only the latter.

This difference is more significant than it may appear to be at first sight. More is known about the "symmetric" version of the Rosser variant than is known about the non-symmetric original version. For example, it is known that the symmetric version violates the F-LPC, but it is not known whether the non-symmetric version does. We became attuned to these points by reading the highly illuminating discussion of Kreisel and Takeuti [1974] (see, in particular, the section on Rosser variants of pp. 46—8).

[37] If T is consistent and closed under cut and Σ_1^0-complete, then it follows that T_R is too. For T_R has exactly the same set of theorems and proofs as T, if T is consistent. Thus, there is, in general, no reason to suppose that Rosser variants will violate the substantive claim of the Kreisel—Takeuti proposal. It's just that they themselves will not be able to prove their adherence to this substantive demand.

[38] This follows from the fact that the procedure for determining whether one formula either denies or is denied by another is effective, and the fact that any element in the given omega-ordering of T's proofs, upon which T_R is predicated, has only finitely many predecessors.

[39] An example of such a constraint is that presented in Feferman [1960]. There Feferman defines a variant system of arithmetic T_F by adding a condition to the usual definition of axiomhood for T. In order to be an axiom of T_F, a formula not only has to pass the test for axiomhood in T, but must also be *consistent with* those axioms preceding it in a given omega-ordering of the axioms of T. Since, by Church's Theorem, there is no effective test for the consistency of quantificational formulae, it follows that T_F cannot be a *formal* system (in the usual sense of the term) because it lacks an effective test for axiomhood (and, hence, for proofhood).

[40] There are two important questions concerning the instrumentalist's use of revision procedures that we should like to leave with the reader. The first concerns the rationality of continued use of extrinsic procedures in the fact of recurrent inconsistency. Specifically, is there a point at which even the *instrumentalist* would be forced, on pain of irrationality, to abandon use of extrinsic procedures in favor of instrinsic procedures? And the second concerns the rationality of employing any revision procedure when there is reckoned to be but little chance of inconsistency in the base theory. In other words, if an instrumentalist believes that there is little chance of T's being inconsistent, is there sufficient point for him to construct his ideal system after the manner of

T_R? And, if he believes that there is significant chance of inconsistency, should he settle for only an extrinsic principle of revision?

Any final determination of the reasonableness of Rosser formalization would, it seems to us, have to offer some answer to these questions. I have more to say on these and related questions in an unpublished manuscript entitled "Formalization", available on request.

[41] Expressions of this general point of view may be found in Kreisel [1976], pp. 76–8, and Prawitz [1981], pp. 254–8 and, somewhat less explicity, in Smorynski [1977] and Giaquinto [1983].

[42] This argument is suggested by Smorynski [1977], pp. 846–7, and Giaquinto [1983], p. 124.

[43] For the idea of how demonstrable Σ_1^0-completeness figures in this proof, see Smorynski [1977], Theorem 4.1.4. Also, see pp. 829–30 of the same paper for some pertinent restrictions on T.

[44] The statement in the 1972 paper appears to contain something like a typographical error; namely, the first occurrence of the term 'consistency' in the fifth line from the foot. It seems to me that Prawitz should want to use 'real-soundness' (my term) rather than 'consistency'.

[45] There are, of course, many (unspoken) assumptions other than (i) and (ii) that are used in this argument (e.g., that the denial of a true Σ_1^0 formula is equivalent to a false Π_1^0 formula, and that the logic of T is the ordinary classical logic). Also, the argument is, in our opinion, problematic (for reasons that shall shortly become clear). Thus, we are not putting it forth as a good argument, but only as a reconstruction of how the Prawitzian would locate the appeal to Σ_1^0-completeness in his derivation of soundness from consistency.

[46] Prawitz [1981], p. 257 notes some of the debilities of '$\mathrm{Prov}_T(\ulcorner A \urcorner) \supset A$' as an expression of T's soundness. But he too fails to adequately mark the difference between finitary and classical truth. Thus, he takes the real-soundness of T to be expressed by the following proposition (where 'g' is the Gödel-numbering function, 'Pr' stands for the "proof of" relation for T, and 'Tr' for the (classical) truth property).

> For each proof p of T and for each real sentence A in T: $\mathrm{Pr}(g(p), g(A)) \supset \mathrm{Tr}(g(A))$.

He then notes that this principle entails each instance of Π_1^0-reflection.

We feel that it is more nearly accurate to say that the real-soundness of T is expressed by the following claim (where 'Prov_T' stands for provability in T, and 'Prov_F' for provability in a system F taken to formalize finitary reasoning).

> For each real sentence A of T: $\vdash_F \mathrm{Prov}_T(\ulcorner A \urcorner) \supset \mathrm{Prov}_F(\ulcorner A \urcorner)$.

Given what we know about finitary reasoning, it seems that we have no right to expect this soundness principle to be equivalent to that suggested by Prawitz. In particular, F's being a codification of finitary reasoning would appear to give us no reason to suppose that the soundness principle just stated implies Prawitz's. Hence, we cannot infer from the premise that demonstrable Σ_1^0-completeness is necessary for the proof that $\mathrm{Con}(T) \vdash_F \mathrm{Prov}_T(\ulcorner A \urcorner) \supset A$, that it is also necessary for a proof that $\text{Con}(T) \vdash_F \text{Prov}_T(\ulcorner A \urcorner) \supset \text{Prov}_F(\ulcorner A \urcorner)$. And so, the motivation that the first defense of Claim 2 supplies for demonstrable Σ_1^0-completeness is, to our mind, unconvincing.

[47] But does it, in effect, oblige him to give up the search for a finitary consistency proof for T as pointless? After all, to give up demonstrable Σ_1^0-completeness is to give up any assurance of getting a finitary proof of T's real-soundness from a finitary proof of its consistency. Thus, what would be the point of having a finitary proof of T's consistency even if one were available? There would, I think, be this much point: since T cannot be real-sound if it is inconsistent, proving its consistency removes one potential obstacle to its real-soundness. Thus, a proof of T's consistency isn't entirely worthless to the Hilbertian even if he cannot derive a proof of T's real-soundness from it.

[48] To avoid misunderstanding, let us now register some disclaimers concerning our use of Rosser variants. First, we are not claiming that they show that Hilbert's Program *can* be carried out. The Rosser consistency formula is provable in the system, but from this it does not follow that real-soundness is. For one thing, Rosser consistency only expresses freedom from "explicit" contradiction (i.e., absence of pairs of theorems one of which is the direct denial of the other). So it does not guarantee the elimination of "tacit" contradictions. Furthermore, something like demonstrable Σ_1^0-completeness appears to be needed in order to infer soundness from consistency, and Rosser provability violates that condition. So, the provability of Rosser consistency doesn't guarantee the provability of real-soundness. (However, since "explicit" incon-sistency is one way in which soundness can fail, its removal brings one a step closer to soundness. So, Rosser consistency needn't be regarded as entirely worthless by the instrumentalist.)

Secondly, we are not claiming that the discovery of a new G2-type result extending to Rosser systems is impossible. Perhaps it can be shown that all of the more "serious" notions of consistency even for Rosser systems are bound by a G2-type result (which would assume, of course, that Rosser consistency is not "serious" even though it makes a "contribution toward" soundness). But such a finding would represent a significant departure from the present generalized version of G2 (i.e., the one based on the Derivability Conditions). For it would at the very least require giving an analysis of what should count as a "serious" notion of consistency for an instrumentalist (a task which is complicated by the fact that a notion of consistency need not entail soundness in order to "contribute toward" it, and thus potentially qualify as "serious" on that count), and a subsequent demonstration that none of the "serious" notions pertaining to Rosser systems is provable. Thus, one would have to tie the unprovability of consistency directly to the ability of a formula to express a "serious" notion of consistency, rather than to the reasonableness of the notion of provability from which the given consistency formula is defined. And this, it seems, would set it apart from the present generalization of G2 which *does* seek to tie the unprova-bility of a consistency formula to the reasonableness of its underlying notion of provability. The SA makes crucial use of the claim that every reasonable notion of provability must satisfy the Derivability Conditions. For it is from this claim that it derives the unprovability of every reasonable consistency formula (via the assumption that a consistency formula is reasonable only if its underlying provability formula is). Our use of Rosser variants is intended to oppose this basic claim by showing that Rosser provability (which violates the Derivability Conditions) *is* an instrumentalistically reasonable notion of provability. In other words, our use of Rosser variants is designed to show that the anti-Hilbertian who employs the current version of G2 cannot solve the Stability Problem. And this is all that we claim for it.

CHAPTER V

THE CONVERGENCE PROBLEM AND THE
PROBLEM OF STRICT INSTRUMENTALISM

1. INTRODUCTION

The main issue of this chapter concerns how the epistemic benefits of an ideal system may come to be distributed over it. In particular, we are interested in whether they are evenly or unevenly distributed over it; where to say that the epistemic benefits of T are evenly distributed over it is to say that there is no isolable subsystem of T containing all or nearly all of its humanly useful ideal proofs. When the epistemic benefits of a system can be compressed into one of its parts, we say that the system is "localizable".

We believe in localizability. That is, we believe it to be a widespread phenomenon among ideal systems. The human epistemic agent operates under the constraint of finite bounds on both the amount of time and effort he can expend on epistemic pursuits, and on his ability to epistemically manage size and complexity. And such limitations dictate that he use only a finite portion of the potential resources (i.e., the ideal proofs) available in any system of ideal proofs. Thus the epistemic benefits of an ideal system are localized in principle by the very finitude of human cognitive capacity.

All this merely recapitulates the Thesis of Strict Instrumentalism (given in Chapter III), and is intended to reflect the practical limitations governing human use of ideal proofs as epistemic instruments.[1] These limitations, in turn, determine a certain localization of any given system of ideal proofs; namely, that finite portion of it that is restricted to ideal proofs having humanly manageable length and complexity. Yet the Hilbertian is not in a position to derive much benefit from such a localization. For the only apparent way of proving its soundness (as opposed to that of some infinite system in which it might be embedded) is to go through it case by case, proving the real truth of the conclusion of each of its proofs. And such a proof of soundness could not be used as the basis for any *gainful* application of the Replacement Strategy, since it would require execution of the very contentual task that use of the Replacement Strategy is intended to avoid. In proving the real-soundness of each proof in a case-by-case manner, one would be forced to give a real proof of each theorem directly; and this is tantamount to working in the real rather than the ideal system. Thus, there would be no epistemic gain in an application of the Replacement Strategy based on a case-by-case proof of real-soundness, and we must conclude that the crude localization suggested by the TSI is not a tenable one for the Hilbertian.

What is needed is some way of embedding the finite set of T's humanly useable proofs in an inductively defined system of proofs whose inductive structure may then be exploited to produce a feasible and efficient proof of soundness, and hence an efficient application of the Replacement Strategy. Thus, ironically, the Hilbertian can utilize the localization suggested by the TSI only by diffusing it.

There are, however, manifest dangers in such a tactic. For what reason is there to believe that the system thus produced will be a *proper subsystem of T*? And even if it turns out to be a proper subsystem of T, what reason is there to believe that it represents a useful localization of T; i.e., one not confronted with all the same obstacles that confront T? These are questions which must be answered if any serious case for localizability is to be made out, and they are, therefore, the questions which we must attempt to answer in the remainder of this chapter.

The argument which results bears on the SA in at least three ways. The first is by means of the so-called Convergence Problem, which brings considerable pressure to bear on premise (3). Even if the usual ideal systems (e.g., PA and ZF) clearly subsume finitary reasoning, not all of their localizations do. Since, therefore, the Hilbertian's interests and obligations are more nearly represented by these localizations, it is no longer clear that his ideal commitments will result in a system that will subsume finitary reasoning. In other words, it is no longer clear that the Hilbertian's ideal interests and the demands of finitary reasoning will "converge" on the same formal systems. Because of this, the truth of premise (3) is not assured; and this remains so even if one adheres to Hilbert's original conception of finitary evidence rather than some more liberal constructivist strengthening of it.

A second way in which our argument bears on the SA is by calling premise (12) (or, more directly, its underlying assumption that every appreciable ideal theory must contain elementary mathematics) into question. Reflection on the conditions governing Hilbertian residues and on the notion of appreciability pertaining to the Hilbertian's ideal theories, clearly reveals a lack of connection between a theory's being appreciable and its subsuming elementary mathematics (in the sense of premise (12)). Thus, the assumption that every appreciable system of ideal proofs must contain elementary number theory is false or, at best, groundless: and, as a consequence, so is premise (12). In what follows we shall refer to this finding as the Problem of Strict Instrumentalism.[2]

More important, in our opinion, than either the Convergence Problem or the Problem of Strict Instrumentalism taken singly is their joint effect of revealing a serious internal tension in the basic strategy of the SA. That strategy (which we shall hereafter refer to as the "Standard Strategy") consists in finding a formal system F such that (a) it is a clear "upper bound" on finitary reasoning, (b) it is a clear "lower bound" on appreciable systems of ideal mathematics, and (c) G2 holds for it and its extensions.[3] Our discussion of the Convergence Problem and the Problem of Strict Instrumentalism reveals that there is no choice of F that doesn't place at least one of (a)−(c) in jeopardy. The secure choice of F with respect to one desideratum is a risky choice of F with respect to another. Thus, the Convergence Problem and the Problem of Strict Instrumentalism not only bring pressure to bear on premises (3) and (12) separately; they also pit these basic elements of the Gödelian's strategy against one another, thereby posing a serious internal dilemma for the SA.

These, then, are the primary ways in which the argument of this chapter is intended to bear on the evaluation of the SA. Since they all depend, in one way or another, on the localizability question for the usual ideal systems (e.g., PA and ZF), it seems appropriate to begin our argument with a more detailed discussion of that notion.

2. LOCALIZATION

It is natural to think of the localization of a system T as involving the isolation of one of its proper subsystems which is then shown to be instrumentally equivalent to it. Moreover, it is natural to think of this subsystem as being formed by taking some logically proper subset[4] of T's axioms, and "closing" it under T's logic (or one of its sub-logics). Conceived of in this way, a localization of T must have not only a strictly narrower class of theorems, but also a strictly narrower class of axioms than T itself.

But though this is the natural way to think of localization, and also the way that we shall usually be thinking of it in this chapter, it is not the only, nor perhaps even the best way to think of it. There are different notions of subsystem that might be employed which do not force a subsystem of T to have a different set of axioms than T. And there are also different conceptions of the *mode* of subsystem formation that can be called upon. In particular, it is possible to distinguish a conditional from an unconditional mode of formation. According to the former, a localization of T might not *actually* call for the elimination of any of T's theorems, but rather make their elimination conditional on some other feature of T (say, its consistency/inconsistency). According to the latter, on the other hand, the localization of T would require the actual discharging of some of T's theorems.

The defense of Rosser formalization in the last chapter provides us with both a model of and a justification for these distinctions. For, given the instrumentalist's tolerance for extrinsic revision procedures, there is no need for his localizations (even his unconditional localizations) to be founded upon a narrowing of the base theory's axioms. They might just as well stem from the elimination of certain combinations of axioms and/or rules of inference. Likewise, given his ability to make use of a revisability ordering of the proofs of an ideal system, there is no evident reason why he should be wed to unconditional rather than conditional modes of localization.[5] His commitment to narrowing the theorems of the base theory need be no stronger than the evidence of its inconsistency, but he should have a stable commitment to eliminating inconsistencies *should* they arise, and to do so by the least costly means (i.e., by dropping the more revisable of a pair of conflicting proofs). Such a commitment has a localizing tendency or intension, since it calls for a restriction of the base theory to the largest consistent initial segment of its revisability ordering. And this being so, it seems to deserve to be classified as a type of localization.

Our interest in the distinction between conditional and unconditional localization is due largely to what it tells us about the Standard Strategy. That strategy depends upon an assumed connection between a system's mathematical power and G2's holding for it; that is to say, if the system is powerful enough to contain the minimum required for appreciable ideal mathematics and/or powerful enough to subsume the results of finitary reasoning, then G2 will hold for it. In the absence of such an assumption, there is simply no reason to believe that the Standard Strategy can be carried out. For if there is no connection between (1) a system's ability to capture finitary thought and/or the minimum required for a significant ideal mathematics, and (2) G2's holding of it, then any intersection of upper bounds on finitary reasoning, lower bounds on appreciable ideal mathematics, and systems for which G2 holds, is purely accidental. And if this is so, the Standard Strategy poses no serious threat to Hilbert's Program, since there should then be ways of doing significant ideal mathematics to which either G2 doesn't apply, or, if it does, does not signify the impossibility of a finitary consistency proof because the ideal mathematics in question fails to subsume finitary thought.

It seems clear, then, that the viability of the Standard Strategy as an anti-Hilbertian tactic depends crucially on an assumed connection between a system's mathematical power and G2's holding for it. But it is just this assumption that the phenomenon of conditional localization calls into question. For conditionally localized theories can avoid G2, even though it may hold for other theories having exactly the same theorems and, therefore, the same mathematical power. This is amply illustrated by the case of Rosser formalization which, as mentioned above, represents one type of conditional localization. Where T is a base system of the usual sort, and T_R its Rosser localization, base theory and localization will be equivalent in their deductive power. Yet whereas G2 may hold for T, it definitely will not hold for T_R. Thus, it becomes clear that there is no positive connection between a system's mathematical power and G2's holding for it.[6]

Conditional localization, therefore, proves to be a powerful antidote to the Standard Strategy. And because of the cogeniality of conditional localization to the instrumentalist conception of ideal theories, it is an antidote that the Hilbertian need have no fear of applying. However, it is not the only defense he has against the onslaught of the Gödelian Challenge. And in the remainder of this chapter, we shall be concerned with showing how there are unconditional localizations of the usual systems in which the Hilbertian may take refuge.[7]

Our plan is as follows. In the next section we shall argue that there are unconditional localizations of the usual systems which force the advocate of the Standard Strategy to modify his estimates of the lower bounds on appreciable ideal theories. And we shall also argue there that this change is problematic, since consistency comes more cheaply for the new estimates then for the old. Finally, in the section following the next, we try to show that this modification of the standard of appreciability for ideal theories prompts a new estimate of the upper bounds on finitary reasoning, and how this, in turn, leads to the Convergence Problem.

3. THE PROBLEM OF STRICT INSTRUMENTALISM

If our case against the Standard Strategy is to have any cogency whatsoever, we must be able to say what is wrong with choosing such obvious candidates as PA as providing both an upper bound on finitary reasoning,[8] and a lower bound on ideal practice. After all, it is commonly accepted that, at least as originally conceived, finitary thought is theorem-wise codifiable in PA. And it is also commonly acknowledged that only a relatively small part of classical mathematics can be done within the confines of PA. So why should the advocate of the Standard Strategy look any farther than PA for a solution to his alleged problems? In fashioning a response to this question, we shall rely upon two things: first, an analysis of what makes an ideal system "appreciable"; secondly, an understanding of what the doctrine of Hilbertian residue tells us about the comparative appreciability of a non-finitely axiomatizable system (such as PA) and certain of its finitely axiomatizable subsystems. We shall eventually combine these elements to show that PA is not a viable choice of theories upon which to found the Standard Strategy.

To begin with, then, let us consider the notion of "appreciability" as it occurs in the Hilbertian's conception of an ideal theory. The key here is to realize that, from an instrumentalist's point of view, the appreciability of a theory is not determined by its mathematical or deductive power *per se*, but rather by its utility as an epistemic instrument (i.e., its utility as a non-literal calculary device for deriving literal truths). Hence, a system T_1 may be deductively weaker than a system T_2 without being instrumentally inferior to (i.e., less appreciable than) it. For if the epistemic yield of T_1 (i.e., the class of literal truths produced by the epistemically useable derivations of T_1) is as great as that of T_2, then T_1 will be just as valuable an epistemic instrument (and, hence, just as appreciable) as T_2, even if T_2 is deductively superior to it. The instrumental appreciability of a theory is, therefore, determined not by the size of its deductive product (i.e., its mathematical power), but rather by the size of its *epistemically useful* deductive product.[9] And the two are by no means the same.

Given this understanding of the notion of appreciability, it is possible to argue that the Hilbertian residue of a theory T is just as appreciable a body of ideal thought as T itself. As inspection of the conditions on Hilbertian residues reveals, the only proofs of T that are deleted in forming T_H are those that add nothing to the epistemic utility of T. For the proofs deleted are either (1) too long and complex to be put to any epistemic use (cf. condition (i)), (2) not ideal proofs at all, and so not even potential contributors to T's *instrumental* epistemic utility (cf. condition (iii)), or (3) represent no gain in epistemic efficiency over the corresponding real methods (cf. condition (ii)). In each case, the proof deleted adds nothing to the instrumental utility of the ideal system, and so dropping it does not, in and of itself, betoken a decrease in instrumental value.

Of course, dropping isolated proofs from a theory might result in a loss of systematicity and, hence, a loss of ability to obtain a systematic (as opposed to a case-by-case) solution to the soundness problem. And if this were so, the class of ideal proofs which one would obtain by performing the deletions called for in conditions (i)−(iii) might actually have less utility than the base theory due to the lack of any suitably efficient way of addressing the soundness question. For, it must be remembered, the viability of the Replacement Strategy depends crucially upon the existence of a systematic solution to the soundness problem. Otherwise, an M-proof of a given real statement R will require an O-proof of R, and so could not possibly claim to represent a streamlining of the O-methods.[10]

This concern for a systematic solution to the soundness problem is, however, addressed in that step in the construction of an Hilbertian residue where we are told to collect the axioms of T that appear in the proofs surviving application of (i)−(iii), and to close this set under T's logic. Because of this feature of their construction, Hilbertian residues possess at least the rudiments necessary to a systematic solution of their soundness problem.

Since, therefore, T_H contains every instrumentally useful proof contained in T, and since its soundness problem (like T's) admits of a systematic solution, and since, finally, a soundness proof for

T_H will not require evidence of a type less trustworthy than that required for the proof of T's soundness, it seems plausible to say that T_H is just as valuable an epistemic instrument as T. Thus, T_H is to be regarded as equi-appreciable with T and, as we shall shortly see, this has serious consequences for the choice of PA as a basis for the Standard Strategy.

But the equi-appreciability of PA_H and PA does not, in and of itself, show that PA_H is a localization of PA. Hence, it does not force the advocate of the Standard Strategy to modify his estimate of PA as a lower bound on appreciable ideal mathematics. In order to obtain such a result, we require, in addition to the equi-appreciability of PA_H and PA, an argument to the effect that PA_H is a *proper* subsystem of PA. For without such an argument, it remains possible that the axioms upon which PA_H is based also form an axiomatic basis for PA. And this would destroy any claim that PA_H might have to being an unconditional localization of PA.

The well-known proofs of the non-finite axiomatizability of PA, however, furnish us with just such an argument.[11] For, by the TSI, we know that only finitely many proofs of PA will be of a humanly useable length or complexity. And, by virtue of the presence of condition (i), this implies that only finitely many proofs of PA will survive the test of conditions (i)−(iii). Hence, the set of axioms of PA appearing in the proofs surviving (i)−(iii), and thus the axiom-set of PA_H itself, will be finite. So, from the non-finite axiomatizability of PA it follows that PA_H is a *proper* subsystem of PA. And from this and the (previously argued) equi-appreciability of PA and PA_H, it follows that PA_H is an unconditional localization of PA.

In our estimation, this is a result having considerable bearing on the Standard Strategy. For, given that PA_H and PA are equi-appreciable, the Hilbertian instrumentalist is under no obligation to defend the latter if he can instead defend the former. Hence, PA is *not* a "lower bound" on what the Hilbertian must vindicate if his program is to have a chance of working. If PA enjoys no instrumental epistemic advantage over PA_H, then the Hilbertian's duties to defend the epistemic utility represented by PA may be fully discharged by proving the soundness of PA_H. And so, the real question is whether the soundness problem for PA_H is any more amenable to finitary resolution than that for PA.

We believe that there is evidence (albeit indecisive) that this is so. And it comes from the proof of PA's reflexivity given in Mostowski [1952] (cf. Theorem XVIII). Roughly, a system T, based on an infinite set of axioms, is said to be reflexive just in case the consistency of every system based on a finite set of T's axioms is provable in T.[12] Thus, the proof that PA is reflexive implies that the consistency of PA_H (i.e., the consistency of the Hilbertian residue of PA) *is* provable in PA. And since, on the current model for the Standard Strategy (i.e., that which is based upon PA), PA is the identified bound on finitary thought, it follows that the consistency problem for PA_H (which is solvable in PA) is more amenable to finitary treatment than the consistency problem for PA (which is not).

Of course, it does not follow from this that the consistency of PA_H is finitarily provable, since it has not been shown that provability in PA implies finitary provability (even for formulae like $Con(PA_H)$). Neither does it follow that the soundness of PA_H is provable in PA. Hence, the provability in PA of PA_H's consistency most assuredly does *not* guarantee that the soundness of PA_H is finitarily provable. Nonetheless, it signifies some progress in that direction. For the soundness of a theory cannot be finitarily proven if its consistency is not provable in PA. Thus, the soundness problem for PA_H passes one test of finitary solvability that the soundness problem for PA fails (and it doesn't fail any that PA passes). And so there is reason to regard it as more congenial to finitary treatment than the soundness question for PA.

By what we have just said, then, PA is a poor choice of theories upon which to base the Standard Strategy. For while it is acceptable as an upper bound on finitary reasoning, it is not acceptable as a lower bound on ideal reasoning. PA has no instrumental advantage over PA_H. Thus, since the soundness problem for PA_H

may be more tractable than that for PA, there are no grounds for saying that the minimal task facing the Hilbertian is the soundness problem for PA.

The advocate of the Standard Strategy is therefore obliged to adjust his estimate of the Hilbertian's minimal responsibilities downward, so as to make it the soundness problem for PA_H rather than the soundness problem for PA. But, as we shall see in the following section, this is a highly problematic adjustment for him to make.

4. THE CONVERGENCE PROBLEM

The basic problem with a downward adjustment of the lower bound on the ideal method is that it entails a corresponding adjustment of the upper bound on finitary reasoning, and this latter adjustment is difficult to defend. To see that this is so, let us consider what happens to the Standard Strategy when PA_H replaces PA as the estimated lower bound on the ideal method.

In order to keep the Standard Strategy intact in the face of such an adjustment, one's first response might be to retain PA as the estimated upper bound on finitary reasoning, and look for a strengthening of G2 to show that not-$\vdash_{PA}$ Con(PA_H). Such a proposal, if successful, would preserve the integrity of the Standard Strategy by giving it the same basic conclusion; namely, that PA_H's consistency (and, hence, its soundness) is not finitarily provable. However this proposal cannot be accepted, since by Mostowski's result on the reflexivity of PA, $\vdash_{PA}$ Con(PA_H).

However, the failure of this proposal is instructive in that it shows the need to replace PA with a lower estimate on the bounds of finitary reasoning if the Standard Strategy is to be preserved. Accordingly, let us now consider how that might be accomplished.

One suggestion along these lines is to take PA_H as the bound on finitary reasoning, and to prove G2 for PA_H. Such a suggestion is, however, implausible. For though it might well be that G2 holds for PA_H,[13] it is not at all plausible to suppose that PA_H subsumes (in even a theorem-wise manner) finitary thought. And this can be made clear by considering the effects of the conditions which govern the formation of Hilbertian residues.

Each of these conditions exhibits a disposition to eliminate various real theorems of T from membership in T_H. However, the conditions under which (iii) has eliminative force are significantly different from those under which (i) and (ii) do. In order for (iii) to eliminate theorems, T must *proof-wise* contain some finitary reasoning, since (iii) calls for the deletion of a real theorem only if it is the conclusion of a real proof. On the other hand, conditions (i) and (ii), have the power to delete real theorems from T even if T only *theorem-wise* contains finitary thought. This difference gains significance from the fact (cited in note 8) that the advocate of the Standard Strategy only requires an upper bound on finitary thought that *theorem-wise* subsumes it. Thus, if our appeal to Hilbertian residues is to have its optimum effect, we must keep our use of condition (iii) to a minimum and base our case primarily on the eliminative effects of conditons (i) and (ii). With this in mind, let us now consider how the working of (i) and (especially) (ii) eliminates any grounds for expecting that PA_H will subsume finitary reasoning.

The eliminative force of (i) is based on the TSI, and consists in the demand that all unfeasibly long or complex ideal proofs of real formulae be dropped from the finite class of proofs whence the axioms of PA_H are taken. Such deletion is, of course, provisional in character since a theorem dropped on these grounds might be "reinstated" for either of two reasons:[14] first, by dint of its having a feasible ideal proof over and above its unfeasible ones, and secondly, by its unfeasible proofs being restored to PA_H through that clause in its construction which calls for the "closure" of the axioms of PA_H under the rules of inference of PA. Nonetheless, since such reinstatement is not guaranteed, condition (i), by itself, makes PA_H a questionable choice of bounds for finitary reasoning.[15]

Unlike condition (i), the eliminative potential of conditon (ii) is not based upon the TSI or any similar appeal to the finitude of human cognitive capacity. Instead, it makes use of Hilbert's distinction between the problematic and unproblematic reals. That distinction enjoins the Hilbertian instrumentalist to refrain from using an ideal proof to derive a real formula if that formula has an unproblematic real proof. Nor does the Hilbertian recommend using an ideal proof to derive a problematic real unless there is a gain in efficiency realized.[16] He is committed instead to using (and thus to defending) only such ideal proofs as are "progressive" (i.e., yield some significant gain in epistemic efficiency over the real proofs). Condition (ii), therefore, calls for the elimination of all real theorems of PA having either an unproblematic real proof, or a real proof which, though perhaps problematic, is nonetheless no less efficient a means of epistemic acquisition than any of the ideal proofs available in PA.

Of course, the eliminative force of condition (ii), like that of condition (i), is somewhat provisional in character. That is to say, it only results in the actual elimination of real theorems from PA to the extent that there are axioms of PA that appear only in "superfluous" ideal proofs (i.e., ideal proofs that represent no gain in efficiency over their real counterparts). However, since (for all we presently know) there may be such axioms, PA_H is not a choice of bounds for finitary reasoning upon which a convincing application of the Standard Strategy can be based.[17]

So far, then, we have argued that neither PA nor PA_H is a propitious choice of bounds for finitary reasoning: PA because it is too strong (i.e., because it is strong enough to prove the consistency of PA_H, and thus unable to rule a finitary proof of the real-soundness of PA_H out), and PA_H because it is too weak (i.e., because it does not clearly provide for the codification of all useable finitary reasoning). This suggests that perhaps the best bet for the Standard Strategy is to pick as its bound on finitary reasoning a theory having more finitary power than PA_H, but less total deductive power than PA. And the natural candidate for such a selection is PRA, since it is both clearly weaker than PA, and also widely believed to be a formal explication of finitary reasoning. Let us therefore consider what happens to the Standard Strategy when it takes PA_H as the lower bound on ideal reasoning, and PRA as the upper bound on finitary reasoning.[18]

There are, to be sure, advantages to such a move. For by avoiding Mostowski's reflexivity theorem it eliminates the problems associated with PA. And it avoids the problems associated with PA_H because PRA is a more defensible bound on finitary reasoning. Nonetheless, it generates certain problems of its own. In particular, it does not provide for the application of a G2-type result, since it is not clear whether $Con(PA_H)$ is provable in PRA.[19] And without this, the Standard Strategy cannot get off the ground. It is, of course, possible that further elucidation of the exact character of PA_H might reveal that $Con(PA_H)$ is not provable in PRA. But nothing that we now know makes this likely. And so, on balance, it doesn't seem that choosing PRA as the upper bound on finitary reasoning and PA_H as the lower bound on the Hilbertian's commitments is of much help in strengthening the case for the Standard Strategy.

On the basis of the argument in this section, then, we would conclude that the Convergence Problem (i.e., the problem of finding a reasonable estimate of an upper bound on finitary reasoning and a reasonable estimate of a lower bound on the ideal method such that the former subsumes, but does not prove the consistency of, the latter) is a serious problem for the Standard Strategy. And though future refinement of both our conception of finitary reasoning and our understanding of humanly useful ideal reasoning might lead to the reinstatement of the Standard Strategy, such refinement would represent a significant advance over what is currently known.

5. CONCLUSION

As we see it, then, one of the central issues with which future work on Hilbert's Program must concern itself is that of localizability. If the usual systems (e.g., PA and ZF) do not adequately reflect the extent to which the epistemic benefits of ideal thought can be localized, then the traditional assessment of Hilbert's Program is not to be trusted. And the appeal to Hilbertian residues (which provide for the unconditional localization of reflexive systems) and Rosser formalization (which gives an epistemological motivation for conditional localization generally) shows that there is at least a conceptual or theoretical basis for taking such localization seriously.[20]

A telling observation of Takeuti's suggests that there might also be an historical basis. In preparing for his proof that analytic number theory is a conservative extension of PA, he notes that though, in logic, it is easy to formulate arithmetical truths that are not provable in PA, in actual practice one hardly ever encounters such. His explanation of this, which we find persuasive, is as follows.

When we learned to formalize mathematics, the formalization itself was an important but difficult task. Naturally we chose a very strong system so that it was easy to see that everything could be formalized in the system. It is likely that we do not need such a strong system and hence our identification of mathematical practice with a certain strong formal system could be an illusion. (Takeuti [1978], p. 73)

What is attractive to one who is seeking guaranteed thoroughness in his codification of ideal practice will not necessarily be attractive to one whose aim is to isolate the source of its epistemic utility. Hence, given that our current formalizations of ideal thought arose primarily out of a desire for thoroughness, they should not necessarily be trusted to provide the sort of description of ideal practice upon which a reliable evaluation of Hilbert's Program might be based.

Thus, one prescription for future work on Hilbert's Program is that of finding a more fine-grained description of those ideal methods which are (either wholly or chiefly) responsible for its efficiency.[21] To do this may, in turn, require the refinement of our current ways of thinking about complexity; that is, it may require the examination of different notions of complexity in order to determine which of these bear(s) the most convincing connection to human (and machine-assisted) cognition and computational capacities. Hence, a second prescription for future work on Hilbert's Program would be to develop a more convincing complexity metric for our formalizations of ideal reasoning.

In addition to this, there is a need for an analysis of complexity as it applies to finitary reasoning. For the feasibility of a given M-proof depends not only upon the complexity of the ideal reasoning involved, but also upon the complexity of the finitary soundness proof which is used to justify that reasoning. In this chapter we can only claim to have removed certain traditional barriers to the *finitary* defense of the ideal method; namely, those imposed by the standard argument from G2. We have not shown that such a defense can be given.[22] Still less have we shown that it can be *feasibly* given. To do so would require the sort of analysis of the feasibility of finitary reasoning mentioned above, and we have none to offer.

Therefore, we cannot claim to have established Hilbert's Program. Instead, we can only claim to have rescued it from its most powerful and influential adversary (i.e., the SA) and to have revealed some of its heretofore unrecognized philosophical potential. In so doing we hope to have laid a foundation for future work on Hilbert's Program that gives its friends a clearer view of their opportunities, and its foes a sharper sense of their responsibilities.

NOTES

[1] For a discussion concerning theoretical or in-principle limitations, see Gandy [1982] and Dummett [1975].

[2] Since in giving this argument, we rely heavily upon the Thesis of Strict Instrumentalism, it seems only proper that we entitle the ensuing problem for the SA the "Problem of Strict Instrumentalism".

[3] Actually, as we shall see later on in the discussion of the Convergence Problem, there is no inherent need to make the estimated upper bound on finitary reasoning and the estimated lower bound on the ideal method be the same theory. One need only require that the former be a subsystem of the latter, and then prove G2 in the following (strengthened) form: not-$\vdash_{F_1} \mathrm{Con}(F_2)$ (where F_1 is the estimated upper bound on finitary reasoning, and F_2 the estimated lower bound on the ideal method).

[4] By a "logically proper subset" of T's axioms, we mean a subset of T's axioms that does not imply, on T's logic, all of T's axioms.

[5] Part of the reason why the instrumentalist has an affinity for conditional localization is because of his capacity for unconditional localization. This capacity, in turn, is based upon the TSI; i.e., the belief that only a finite portion of the proofs of an ordinary ideal system are efficient enough to be put to any gainful epistemic use. Because he believes this, the thought (which is fundamental to conditional localization) that he might have to make do with a finite part of an ideal system's resources is not an unduly troubling one to the instrumentalist.

[6] There may, however, be a negative connection. That is, there may be systems (e.g., system V of Theorem 1.5 of Jerslow [1976]) that are too weak to sustain G2.

[7] Actually, our argument shall deal with the unconditional localizations of only a subclass of the usual systems; namely, the so-called reflexive systems (e.g., PA and ZF). We shall define the notion of reflexivity shortly.

[8] In calling a formal system an upper bound on finitary thought, the advocate of the Standard Strategy is not committed to saying that the proofs or reasonings of finitary thought are expressed by derivations in the system, but only that the results (i.e., the conclusions or "theorems") of such reasoning are expressed by theorems of the system. This is a result of the fact that in order to refute the Hilbertian's program for a given ideal system S, he need only show that its consistency is not finitarily provable; and in order to do this, he need only supplement a proof of G2 for S with an argument to the effect that its consistency is finitarily provable only if $Con(S)$ is a *theorem* of S.

[9] Strictly speaking, of course, it is not just the size of the useful product that counts, but the *degree* of its usefulness (i.e., the degree of its efficiency and reliability) as well. However, since such considerations are not particularly germane to the present argument, they will not be emphasized.

[10] This just reiterates the by now familiar claim that the Replacement Strategy cannot be based upon an unsystematic (i.e., case-by-case) approach to the soundness problem, since such an approach entails use of the very O-methods that M-replacement is intended to avoid.

[11] For a "direct" proof of this see Ryll-Nardzewski [1952]. And for a proof based upon the reflexivity of PA and an appeal to G2, see Mostowski [1952], or Montague [1957]. By calling PA non-finitely axiomatizable, we mean that the theorem-set of PA is not the logical closure of any finite set of its axioms.

[12] Put less roughly, T (a system based on an infinite number of axioms) is reflexive if and only if for every system T_F based on a finite number of T's axioms, $Con(T_F)$ (i.e., the "usual" formula of T taken to express T_F's consistency) is provable in T.

[13] After all, G2 is proven for Q in Bezboruah and Shepherdson [1976], and it doesn't seem implausible to suppose that PA_H will include Q. At any rate, we won't contest such an assumption.

[14] Of course some of the real theorems eliminated by (i) are eliminated nonprovisionally. This is true, for example, of those real theorems which are

themselves so long and/or complex that any proof containing them must, by the very fact of that containment, be unfeasible. We are, however, not very interested in such cases, since what we really want to know is whether all *feasible* finitary reasoning is codifiable in PA_H; and the elimination of theorems which are themselves unfeasible still leaves this possibility open.

So, we are interested in condition (i) because of its capacity to potentially eliminate feasibly complicated real theorems. But why? The answer, once again, is that the Hilbertian is interested only in gainful applications of the Metamathematical Replacement Strategy. And since any such application must be based upon a manageable proof of real-soundness, it follows that unfeasibly complicated real theorems can play no part in carrying out the Hilbertian's program. Thus, the question of crucial interest to him is not whether PA_H subsumes *all* of finitary thought, but whether it subsumes all of *feasible* finitary thought.

[15] Could condition (i) lead to the elimination of anything but real theorems that are *themselves* unfeasible? The answer is 'yes'. For it is at least possible that there are axioms of PA which appear only in unfeasible proofs of feasible theorems (i.e., theorems which are not by themselves so complicated as to be unfeasible). Still, we do not believe that the eliminations authorized by condition (i) cast nearly so much suspicion on PA_H's ability to serve as a bound on feasible finitary proof as those authorized by condition (ii). Therefore, most of our case against PA_H is to be seen as resting upon condition (ii).

[16] In other words, even if R is a real formula having *only* problematic real proofs, it does not necessarily follow that the Hilbertian will back some ideal proof of R. For if the ideal proofs of R are themselves long and complex enough, there may be no gain in replacing the problematic O-proofs of R with an M-proof.

[17] It should be noted that the real theorems of PA potentially eliminated by condition (ii) need not be unfeasible. In fact they might be highly feasible and provable by highly feasible ideal proofs. This is so because a real theorem's being provable by means of a highly feasible ideal proof does not prohibit it's also being provable by means of a highly feasible real proof. Thus, condition (ii) targets a different class of eliminations than condition (i).

[18] We do not believe that there is any point to considering the case where PRA is taken as a lower bound on ideal reasoning. For inasmuch as PRA is taken to be a formalization of finitary reasoning, the Hilbertian is neither obliged to nor interested in proving its soundness.

[19] Note that such a result cannot be derived from G2 for PA_H since there are no grounds for believing that PRA is a subsystem of PA_H.

[20] Of course, the unconditional localization associated with Hilbertian residues, and the conditional localization associated with Rosser formalization have quite different bearings on the traditional assessment of Hilbert's Program. The former suggests that the usual formalizations of ideal mathematics are too strong, and that if we weaken them appropriately, we either lose G2 for the weakened system, or the Convergence Problem becomes unsolvable (i.e., the weakened system doesn't subsume finitary reasoning, so that even if G2 *does* hold for it, no anti-Hilbertian conclusion ensues). The latter, on the other hand,

attacks the very idea of there *being* a connection between the mathematical strength of a formal system, and G2's holding of it. And since belief in such a connection is so deeply embedded in the Standard Strategy, it follows that the challenge raised by Rosser formalization penetrates to the very heart of the Gödelian's tactics.

[21] It is worth noting that the Hilbertian could claim a large degree of success for his program even if he couldn't carry it out for the full range of useful ideal methods. Thus, if he could satisfactorily establish the reliability only of that part of ideal mathematics that is responsible for *most* of its efficiency, he could claim a large measure of success for his program. Hilbertian instrumentalism therefore appears to have considerable built-in resiliency; so long as there is *some* class of O-proofs whose replacement by M-proofs yields a significant epistemic gain, the Replacement Strategy (i.e., Hilbert's Program) will admit of significant application.

[22] That is, we have not shown that the real-soundness of our localizations of the usual systems (e.g., PA_H and PA_R) *can* be proven finitarily.

HILBERT'S PROGRAM AND
THE FIRST THEOREM

The argument of Chapters IV and V was designed to show that the typical anti-Hilbertian argument based on G2 is fallacious. There are, however, those who believe that the traditional argument from G2 is not the only, nor perhaps even the best, counter to Hilbert's Program. In the opinion of these writers, an application of G1 can carry the day against the Hilbertian even if the usual argument from G2 should fail to do so. It is these purportedly anti-Hilbertian uses of G1 that we shall consider in this appendix.

There are two interestingly different arguments from G1 that are worth discussing. The first of these is stated by Smorynski in his excellent survey of the incompleteness theorems. After stating G1 and G2 he writes:

> The Second Theorem clearly destroys the Consistency Program. For if R cannot prove its own consistency, how can it hope to prove the consistency of I? . . . Even the First Theorem does this since (1) the statement φ is real; and (2) φ is easily seen to be true. ((1) requires looking at the construction of φ; (2) is seen by observing that φ asserts its unprovability and is indeed unprovable.) Thus, the First Theorem shows that the Conservation Program cannot be carried out and, hence, that the same must hold for the Consistency Program. (Smorynski [1977], p. 825)[1]

Smorynski's argument is ambiguous. He claims that the existence of an unprovable but true real formula is somehow supposed to imply the failure of Hilbert's Program. Yet he doesn't tell us whether it is the unprovability in R or the unprovability in I of this formula that is supposed to supply the anti-Hilbertian thrust of G1. We assume, of course, that I and R many be so related as to guarantee that if φ is unprovable in I, it will also be unprovable in R. So, the ambiguity with which we are concerned is not that pertaining to whether we should take Smorynski as asserting that

φ is unprovable in R or that it is unprovable in I; we assume that he means both. Rather it is that concerning which of these (i.e., the alleged real-incompleteness of R or that of I) is supposed to invest G1 with its anti-Hilbertian force. We shall resolve this ambiguity strategically by arguing that neither alternative has any merit.

Let us consider, then, the argument from the claim that there is a true real formula that is not provable in R, to the conclusion that Hilbert's Program can't work. Here, the chief task facing us is that of figuring out what sense there might be to saying that there is a true real formula φ that is not provable in R, given that R is said to be a codification of real reasoning. If truth is understood finitarily, then a real formula φ will be true only if it is finitarily provable. Thus, the only way that there could be a true real formula that is not provable in R is if not all real reasoning is codified in R.

As we see it, this could happen for either of two possible reasons: (i) R itself is not a complete formalization of finitary reasoning, though there might be such, and (ii) R is not a complete formalization of finitary reasoning, and nothing that we would want to call a formal system could be. If the first of these two alternatives gives the reason why φ is not provable in R, then surely the unprovability of φ in R cannot be used to do the Hilbertian in, since there is nothing in his position that commits him to such a preposterous claim as that every formal system which codifies any finitary reasoning codifies all of it. So, if anti-Hilbertian force is to be found in the claim that there is a finitarily true real formula that is not provable in R, it must arise from the fact that this is necessarily true of all R. Let us assume, then, that this is the case; that is, that the claim which is supposed to have anti-Hilbertian force is that which says that because of G1 there is no formal system capable of codifying all finitary proofs. Is such a claim plausible? And if so, does it have anti-Hilbertian consequences?

The answer to the first question, it seems to me, is 'no'. There is simply no reason to say that the undecidable formula φ produced by G1 is finitarily true (i.e., finitarily provable). The only argument for its truth is one designed to show its classical and not its finitary truth. This argument (which is alluded to in clause (2) of Smorynski's remark) may be stated (roughly) as follows:

> φ says of itself that it is unprovable in R.
> It is true that φ is unprovable in R.
> φ is true.

This clearly is not a finitary proof of φ's truth, however, since its second premise can only be proven finitarily if the consistency of R can be, and that is supposed (by the anti-Hilbertian) to be impossible.[2] Thus, the usual argument for the truth of φ is an argument for its classical truth, and not a finitary proof of φ. As such, *it* provides no reason for believing that there is a finitary truth that is not provable in R.[3]

However, the Hilbertian need not maintain the finitary unprovability of φ in order to withstand the current attack on his position. For even supposing that φ *is* to be taken as a finitary truth, the fact that it is not provable in R does not necessarily promise dire consequences for the Hilbertian. This is so because the unprovability in R of φ (supposed now to be a finitary truth) would be a problem for the Hilbertian only if it (i.e., φ) turned out to be provable in one of the *ideal* systems that he is committed to defending.[4] Thus, unless he is committed to defending the soundness of an ideal system in which φ is provable, the unprovability of φ in R will be of no consequence to him. In order, then, to give a convincing anti-Hilbertian argument from G1 as applied to R, one would have to establish not only that (1) φ is a finitary (as opposed to a classical) truth, and that (2) there is no formal system expressing all and only finitary proofs, but also that (3) there is some ideal system proving φ to whose defense the Hilbertian is committed.

Here we shall only register our disagreement with (3) and note that when it re-emerges at a later juncture, we shall present a more detailed case against it. For the time being, however, we don't need a case against either (3), (2) or (1) in order to make our point; which is that G1 as applied to R lacks anti-Hilbertian force. The mere unprovability of φ in R cannot reasonably be taken as impugning the Conservation Program for I (which is what Smorynski takes the anti-Hilbertian impact of G1 to be) even if (1), (2) and (3) are accepted. The Conservation Program for I fails only if I proves some real formula not provable in R. But φ is clearly not an example of such (even if it is taken to be a true real formula) since it is no more provable in I than in R. Hence the existence of φ does not imply that there is a real formula provable in I that is not provable in R, and thus does not imply the failure of the Conservation Program for I.

Let us consider, then, the other way of reading Smorynski; that is, as attributing the anti-Hilbertian force of G1 to the fact that it shows the existence of a true real formula φ that is not provable in I (= a given system of ideal mathematics). His precise claim (cf. the passage quoted earlier) is that G1 is quite as fit an implement for the destruction of Hilbert's Program as G2. Like G2, it is supposed to extinguish any reasonable hope of obtaining a finitary proof of either the real-soundness or the consistency of I.[5] The only difference between the two results, in Smorynski's view, is that G2 directly attacks the search for a finitary consistency proof (and so, by implication, the search for a finitary proof of real-soundness), while G1 directly attacks the search for a proof of real-soundness (which, in the company of a suitable proof of the real-completeness of I, amounts to an attack on the search for a consistency proof as well). Thus, Smorynski's chief claim regarding G1 is that in proving the existence of a true real formula φ that is not provable in I, it shows that the real-soundness of I cannot be finitarily proven.[6] We shall now attempt to expose the implausibility of this claim.

Let us begin by noting, once again, that if the claim that φ is a true real formula is to have any credence at all, it must be seen as employing a non-finitary notion of truth.[7] But considerations of plausibility aside, such an assumption is now strategically necessary. For if φ were presumed to be a finitary truth, it could not possibly be used to counter the Conservation Program for I, since the Conservation Program is concerned with real formulae that *are not* finitarily true rather than with those that *are* finitarily true. That is, in order to run contrary to the Conservation Program for I, a real formula must be provable in I but *not* finitarily provable. Hence, if φ were taken to be finitarily true, it would automatically forfeit any ability to oppose the Conservation Program for I.

It therefore follows that to be even minimally fit for the anti-Hilbertian application envisaged by Smorynski, φ would have to be taken as a non-finitary rather than a finitary truth. But even supposing that it is to be regarded in this way, its significance for the Conservation Program for I is far from clear. Indeed, its *in*significance seems clear! For, to repeat the theme sounded above, the Conservation Program for I demands only that every real formula provable in I also be provable in R. Since, therefore, φ is, by hypothesis, *not* provable in I, it is clearly unsuited for use as a counter-example to the claims of the Conservation Program for I. And as a result of this, we are led to conclude that the current interpretation of Smorynski's claims fails to yield any genuinely anti-Hilbertian reading of G1. Our final verdict regarding Smorynski's argument from G1 is, therefore, that regardless of whether one reads it as saying that it is φ's unprovability in R or its unprovability in I that has untoward consequences for Hilbert's Program, it must be deemed a failure.

The reader might, however, respond by saying that we have failed to appreciate the real anti-Hilbertian force of G1 because we have mistakenly interpreted it as concerned with the Conservation Program *for* I, rather than with the Conservation Program conceived more broadly. In other words, the unprovability of φ in I is *not* to be taken as jeopardizing the Conservation Program *for* I, but rather the Conservation Program for ideal methods considered as a whole. The argument underlying this position would presumably run as follows.

> For any ideal system I whatsoever, there *is* an ideal proof of φ (viz., the non-finitary argument establishing its intuitive truth).
>
> Moreover, φ is a real formula, and it is not finitarily provable (else it would be provable in R, and hence in I).
>
> But if there will always be a real formula that is ideally but not finitarily provable, then the Conservation Program can never be carried out with finality.

The Conservation Program can never be carried out with finality.

Now we don't see any way of interpreting Smorynski as making this argument, so we won't attribute it to him.[8] But there are others who seem to have had something like this in mind. In their opinion, Hilbert's aim was to provide a "final" proof of real-soundness; i.e., a proof that would provide certainty that every possible system of ideal reasoning is real-sound. Expressions of this point of view occur at various places in Kreisel's writings, among which perhaps the clearest and most informative is that where he describes Hilbert as aiming to give

... a *final solution* of all foundational questions by purely mathematical means, specifically, by a general method for deciding whether or not *any given arbitrary formal system is consistent.* He was convinced that the notion of formal system was sharp enough for mathematical analysis (as was later verified by Turing). Such a method would provide a final solution, at least 'in principle': we should know, here and now, how to decide the consistency of any formal rules we encounter, whatever their source (semantic analysis or formal experimentation). (Kreisel [1976], pp. 111–12)[9]

And, citing Kreisel with apparent approval, Prawitz writes,

It may be said that the whole idea of securing transfinite [= ideal, in our terminology] reasoning or reducing it to finitary reasoning ... now becomes less convincing when this cannot be done once and for all; it may seem that one needs insights into the transfinite in order to know how to extend a given codification. And there will always remain transfinite principles that are intuitively correct (from a transfinite point of view) but have not been secured or reduced. One could try to defend the program by noting that the studied theories are complete in the empirical sense of codifying the actual, existing mathematical practice. Kreisel's point is then that mathematical practice, if that is what one is interested in, can be codified by other systems that use less strong transfinite principles. Hence, in any case, the discovery of incompleteness must lead to greater attention to the choice of codifications. (Prawitz [1981], p. 263, brackets mine)

We would agree with this last claim of Prawitz' that Gödel's Theorems really do force the Hilbertian to pay closer attention to the choice of codifications for ideal mathematics. Indeed, that was the leading theme of the argument of Chapter V. And though it would be nice for the Hilbertian to be able to be very liberal in his estimates of the ideal method, it hardly seems a serious drawback to his program that there are bounds to such liberality. At any rate, our impression is that this is not what most logicians and philosophers of mathematics who are critics of Hilbert's Program see as its major weakness. The prevailing attitude seems rather to be that the only thing that could possibly save Hilbert's Program is the discovery of some finitary reasoning that is not codifiable in the usual formal systems of ideal mathematics. And that is generally viewed as virtually, if not *in principle* impossible.

We shall not, however, debate this issue concerning the certainty of our estimates of finitary reasoning. Rather, the chief source of our disagreement is the claim that Hilbertian instrumentalism is committed to obtaining a "final solution" to the problem of foundations for ideal mathematics. More particularly, we find no reason to saddle the Hilbertian with the hopeless task of finding a "purely mathematical" defense of the ideal method. But in order to present our case clearly, we need to distinguish some of the rather different aims that might come to be lumped together under the general rubric of a "final" or "purely mathematical" foundation for ideal mathematics.

To begin with, then, let us distinguish two types of "finality"; one having to do with the *certainty* of the Hilbertian's defense of the ideal methods, the other with its *generality*. According to the one conception, the Hilbertian is to be seen as committed to finding a demonstration of the real-soundness of his ideal methods that possesses *final* or *absolute* certainty. And according to the other, he is to be taken as seeking an *a priori* or *mathematical* guarantee that his demonstration encompasses *all* methods that could conceivably be counted as ideal methods; that is, he seeks a mathematical guarantee of the generality of his ideal methods.

Now if the Hilbertian is to be understood in either of these ways, his program must surely be regarded as a failure. For as was argued in Chapter II, there is no reason to regard even a finitary proof of the reliability of some body of ideal methods as being infallible or unsusceptible to revision. This follows from (among other things) the fact that the finitary methods that one would need to use in order to give such a proof are *not* the most certain finitary methods imaginable. Thus, if one were to discover a conflict between these methods and certain other pieces of finitary reasoning, it would not (to say the very least) be a foregone conclusion that they (i.e., the methods used in the proof of reliability), and *not* the others, would survive. And even if, as a contingent matter of fact, the former methods were regarded as the optimally certain cases of finitary reasoning, there would still be no *a priori* or mathematical guarantee of this. Therefore, regardless of one's actual estimate of the relative strength of the finitary evidence to be used in constructing a proof of real-soundness, it seems safe to say that it could not reasonably be claimed to have *a priori* or mathematical finality. And as a result of this, a search for absolute certainty in his proofs of real-soundness would appear to be an entirely unwarranted thing for the Hilbertian to undertake.

Much the same can be said of any version of Hilbertianism that is so ill-conceived as to include among its goals the securing of a mathematical guarantee of its own generality. For whatever else may be the case, it seems clear that one could never be in a position to claim, on *a priori* grounds, that his chosen formalization of a given body of ideal mathematics is the best available, or even that it should always be regarded as minimally adequate. Deciding whether a formal system adequately captures a given body of informal mathematics is always a delicate *a posteriori* matter revolving on such questions as which aspects of the informal reasoning are essential and which inessential to it, what the underlying logic of the given body of reasoning is, what the best choice of postulates is, and so on.[10] I am not saying that such questions are undecidable, nor even that they admit of no answer that is clearly the best among a class of available competitors. Rather, I am only saying that one can have no *a priori* or mathematical warrant for believing that what he takes to be the best (or even a correct) formalization of a given body of informal ideal reasoning really is.

For present purposes, it is important to note that the argument that we have just presented against the possibility of a "final" solution (either in the sense of "final" certainty, or in the sense of "final" generality) to the problem of foundations for the ideal methods makes no reference to G1. So, if the importance of G1 is supposed to reside in the fact that it shows us that the Conservation Program cannot be carried out with absolute certainty and mathematical generality, then it can only be regarded as a largely superfluous and, therefore, rather unimportant result. Let us consider, therefore, whether the Kreisel-Prawitz type of argument can be brought to bear against reasonable formulations of Hilbert's Program (where, by a "reasonable" formulation, we mean one that is not committed to such ends as either a finally certain or a mathematically general execution of the Conservation Program).[11]

The conclusion of the Kreisel–Prawitz argument, as we have presented it, is that the Conservation Program cannot be carried out with finality. If, however, this is to be seen as a conclusion having genuinely anti-Hilbertian import, two things must happen: first, it must be shown that in some sense of the term, the Hilbertian is committed to finding a "final" solution to the con-servation problem; and secondly, it must be shown that G1 forbids finality in precisely that sense. Our contention is that these tasks cannot be satisfied simultaneously. There are senses of "final" in which G1 clearly forbids a final solution to the Hilbertian's problems; and there are senses of "final" according to which it is reasonable to say that the Hilbertian seeks a final resolution of his foundational problems. But these senses of "final" do not overlap, and because of this, we claim that G1 does no intrinsic harm to the Hilbertian's project.

Let us turn our attention , then, to clarifying the sense or senses in which the Hilbertian may reasonably be said to be in search of a final defense of ideal mathematics. Here, it seems to me, the most that can be claimed is that the Hilbertian is under some obligation to given a *lasting* defense of his ideal methods. And in order to do that (for any given area of ideal thought), all that is required is that he be reasonably sure that the ideal methods whose soundness he intends to prove will not be superceded by some other set of ideal methods whose soundness would not be covered by that proof.[12] This, in turn, would require only a reasonable certainty that the system towards which the intended proof of soundness is directed expresses all instrumentally useful proofs pertaining to the area of ideal mathematics that it is supposed to codify.

Failure to attain such certainty would, it seems to me, detract from the desirability of a proposed "solution" to the Hilbertian's foundational problems by leaving its durability in question. The Hilbertian would certainly want to avoid having to redo his soundness proof very often, so the durability of his soundness proofs is something that is important to him. But by the opposite token, a practical assurance of the survival of his foundations for ideal mathematics is seemingly all that he would require. Therefore, unless it can be shown that G1 destroys the Hilbertian's chances of acquiring such practical certainty regarding the compass of his ideal methods, it (i.e., G1) cannot rightly be taken as denying him a "final" solution to his foundational concerns (in the appropriate sense of that term).

The critically important question, therefore, is whether G1 offers any convincing grounds for saying that one can never be in a position to maintain with reasonable (if non-mathematical) certainty that a given system of ideal methods includes all of the instrumentally useful ideal proofs germane to the area of ideal mathematics for which it is to serve as a formalization. Our position is that it does not, and that to believe otherwise is to be guilty of a fallacy; namely, that which reasons from the impossibility of a *mathematical* guarantee that one's formalism is exhaustive of the instrumentally useful ideal methods (of the area of ideal thought that it is intended to formalize) to the impossibility of *any reasonable* guarantee whatsoever (mathematical or non-mathematical) that this is so.

To effectively counter our position, one would have to argue that I's failure to prove φ is always sufficient grounds for either denying or suspending judgement with regard to the claim that I codifies all of the instrumentally useful ideal arguments formulable in its language. And this, in turn, would require an argument to the effect that one cannot possess evidence sufficient to justify believing that ideal proofs involving φ lack significant utility as devices for obtaining knowledge of real mathematics. Put another way, what our objector would have to do is to make a case for saying that there will always be a reason for believing or at least suspecting that $I + \varphi$ is of greater instrumental utility than I (when judged according to the Hilbertian's standards of epistemic utility). We have already registered our disagreement with such a claim,[13] but it remains to be seen whether we can support that disagreement with an argument. This is what we shall now attempt to do.

Hilbert was out to vindicate those methods and results of classical mathematics that make it the intellectually valuable enterprise that it is. In his view, a significant part of this value is due to the efficiency of its ideal apparatus. However, the Hilbertian is not committed to prizing efficiency for its own sake. Like anyone else, he can make discriminations between the importance of different parts of mathematics; regarding some parts of it as more interesting or significant than others. Such discriminations might find their ultimate basis in purely mathematical concerns (which is in keeping with the rather plausible view that not all true propositions of mathematics are of the same degree of *mathematical* interest), or they might come to rest on considerations of empirical applicability (in accordance with the reasonable belief that not all true propositions of mathematics have equal value as empirical or physical applicanda). But whatever their ultimate basis, such discriminations will create an epistemic space of mathematical propositions in which some truths are weighted more heavily than others. And because of this, the value of a given system of ideal mathematics will not be determined simply by its efficiency as a generator of real mathematics. Two ideal systems might generate equally extensive bodies of real truths at the same level of efficiency and still not be instrumentally equivalent if the bodies of real truths thus generated are not of equal interest or importance. Conversely, one system might generate a more extensive body of real truths than another (and do so with no loss of efficiency) and still not be of superior instrumental value because of an overall lack of importance among its "surplus goods".

There are, then, three factors to consider when seeking a comparative estimate of the instrumental value of two ideal theories: (1) the extent of their real products, (2) the efficiency with which these products are generated, and (3) the relative importance of the two products.[14] And bearing these factors in mind, let us now consider the relative instrumental merits of I and $I + \varphi$. [From what we have just said, it should be clear that one might grant that the real product of $I + \varphi$ is more extensive than that of I without thereby conceding that $I + \varphi$ has greater instrumental value than I. Consequently, we shall focus our attention on factors (2) and (3) rather than on factor (1)].

There appears to be no reason to suppose that $I + \varphi$ generates its real product more efficiently than I generates its. Of course, to be certain of this, one would have to enter into a detailed proof-theoretic comparison of the two. But since such considerations form no part of the current anti-Hilbertian argument from G1, and since without them there is not the slightest reason to believe in the superior efficiency of $I + \varphi$, we assume that the advocate of this argument would not wish to be seen as resting his case on such a claim.

Likewise, in the case of factor (3), there is, in general,[15] no reason to regard $I + \varphi$ as having either greater mathematical or greater empirical significance than I. And though, once again, a detailed investigation might turn up some important differences between $I + \varphi$ and I, historically speaking, all the evidence suggests that it would be the exception rather than the rule for the former to possess greater mathematical or empirical significance than the latter. Therefore, it seems safe to say that, in general, $I + \varphi$ fares no better than I with respect to factor (3).

The upshot of all this is that $I + \varphi$ possesses no greater instrumental value than I. And if this is so, then the Hilbertian need not prove the real-soundness of $I + \varphi$ in order to carry out his program; he might do it just as well by proving the conservativeness of I (or one of its *subsystems* that has just as much instrumental value as it!). Moreover, there is no good reason to regard this solution as insufficiently "final". It is *not* mathematically final, since we have no *a priori* guarantee that we will continue to regard a system in the same way tomorrow as we regard it today. But we might have all manner of practical, non-mathematical assurances that the instrumental value of a given formalism will not be superceded by one of its future competitors. And that, it seems to us, is all that we have any right to demand of the Hilbertian.

It is on account of this that the Kreisel–Prawitz argument fails. The Hilbertian's inability to carry out the Conservation Program for $I + \varphi$ generally, shows no weakness in his overall program because he is under no obligation to grant the instrumental superiority of $I + \varphi$ to I (and various of its subsystems) generally. Inasmuch, then, as he can be reasonably certain that $I + \varphi$ represents no meaningful gain in instrumental utility over I, the intractability of the conservation problem for $I + \varphi$ need not be a matter of any special concern to him. And so it is that the present attempt to put G1 to anti-Hilbertian use fails.[16]

Let us close, then, with a brief summary of our argument. We have considered, and rejected, two different attempts to use G1 to refute Hilbert's Program. The first of these (Smorynski's argument) reasoned from the fact that G1 holds for I to the intermediate conclusion that the Conservation Program for I cannot be carried out and, thence, to the final conclusion that Hilbert's Program cannot be carried out. We rejected this argument on the grounds that the existence of an intuitively true real formula (viz., φ) that is not provable in I does *not* imply that the Conservation Program for I cannot be carried out.[17] The second argument we considered (viz., that of the Kreisel–Prawitz variety) sought to link the failure of Hilbert's Program to G1 by pointing out that the general applicability of G1 to the Hilbertian's ideal theories implies that there will always be a way of extending those theories in such a way as to guarantee the ideal provability of a real formula that is *not* finitarily provable. This is a deeper and more sophisticated argument than Smorynski's, but it too is unacceptable, since it is based on the unfounded assumption that the Hilbertian must regard the extension $I + \varphi$ of I as having greater instrumental value than I itself. Overall, then, our finding has been that there is no basis for claiming that G1 refutes Hilbert's Program.

NOTES

[1] In this passage 'R' is taken to stand for a formal system that codifies real or contentual thought, 'I' a formal system of ideal reasoning, and 'φ' the undecidable formula of G1.

[2] If it is not impossible, then, Smorynski is wrong to say, as he does, that G2 clearly destroys the Consistency Program.

[3] If, therefore, Smorynski's position is that the argument suggested by his clause (2) yields a demonstration of φ's finitary truth, then it would seem that he has simply conflated the finitary and classical notions of truth.

[4] In such circumstances, there would be a problem of sorts for the Hilbertian. It would not have been shown that there is an ideally provable real formula that is not finitarily provable, but there would be an obstacle to showing *finitarily* that every real formula with an ideal proof also has a finitary proof. For giving a finitary demonstration of this requires having a formalization of both the ideal and the finitary proofs of a given real formula, and showing that there is a purely syntactical way of transforming the one into the other.

But there simply is no scheme of finitary operations capable of transforming a syntactical object like an ideal proof (or, rather, a formalized ideal proof) into an epistemologically genuine object like a contentual proof. Hence, it seems that the Hilbertian's proposed defense of the ideal method depends upon formalizing both the real and the ideal methods of proof. However, the Hilbertian does not necessarily require a complete formalization of either the ideal or the real methods, since, strictly speaking, he only requires a defense of the useful ideal methods, and hence only a formalization of those real proofs that prove theorems for which there is also a gainful ideal proof. Thus since not every ideal proof is useful, nor every real proof a proof of a theorem possessing a gainful ideal proof, it follows (at least in principle) that neither I nor R need be complete. We will have more to say on this general subject later on in this appendix.

[5] The Hilbertian's search for finitary proofs of the real-soundness and consistency of I are termed (respectively) the "Conservation Program" and the "Consistency Program" by Smorynski. Throughout this appendix we shall follow Smorynski's terminology.

[6] In his terms, the claim is that G1 implies the failure of the Consistency Program. He also makes the associated claim that the failure of the Conservation Program implies the failure of the Consistency Program. But while (for reasons indicated in Chapter IV) we believe that this is debatable, we shall not pause to develop such a challenge here. The chief reason for not doing so is that saving the Consistency Program would be small comfort to the Hilbertian if he had to concede the loss of the Conservation Program since it is the Conservation Program (i.e., the proof of the real-soundness of his ideal methods) that he *really* needs in order to overcome Frege's Problem and the Dilution Problem. In our opinion, the Hilbertian's only reason for being interested in consistency proofs at all is the contribution that they make to the Conservation Program: they always remove one potential obstacle (viz., inconsistency) to real-soundness, and they might (depending upon the availability of appropriate real-completeness results for the given ideal system) be supplementable in such a way as to yield a proof of real-soundness.

[7] The non-finitary argument for its truth goes roughly as follows. φ can be identified with a sentence '$(x) \sim \mathrm{Prf}_I(x, \ulcorner\varphi\urcorner)$' (where '$\mathrm{Prf}_I(x, y)$' "expresses" the idea that the sequence of formulae having Gödel number x is a proof-in-I of the formula having Gödel number y) which "says" that φ is not provable in I. Moreover, one can show that for each numeral $\bar{n}$, $\vdash_I \sim \mathrm{Prf}_I(\bar{n}, \ulcorner\varphi\urcorner)$. Furthermore, nothing is lost by letting I be a system that is (classically speaking sound with respect to its local pronouncements about primitive recursive relations such as '$\sim \mathrm{Prf}_I(x, y)$'. Thus, since I may be assumed to be sound with respect to its theorems of the form '$\sim \mathrm{Prf}_I(\bar{n}, \bar{m})$', and since each instance of '$\sim \mathrm{Prf}_I(\bar{n}, \ulcorner\varphi\urcorner)$' is a theorem of I, it follows (by the assumption that everything in the domain of the standard interpretation of I is designated by some numeral) that '$(x) \sim \mathrm{Prf}_I(x, \ulcorner\varphi\urcorner)$' (i.e., φ itself) is true.

[8] Our reason for saying this derives from the fact that in the passage quoted earlier, Smorynski backs his assertion that G2 destroys the Consistency Program with the claim that since (via G2) R cannot prove its own consistency, it clearly cannot prove the consistency *of I*. Thus, by the "Consistency Program", Smorynski appears to mean the Consistency Program *for I* (where I is any formalization of ideal reasoning extending R).

He next claims that G1 does the *very same thing* (i.e., destroys the Consistency Program *for I*). And the reason he offers for this is that G1 destroys the Conservation Program. But the only Conservation Program whose failure could (with the help of the appropriate proof of real-completeness) imply the failure of the Consistency Program for *I* is the Conservation Program *for I*. Thus, in order to make sense of Smorynski's claim that G1 (like G2) refutes the Consistency Program for *I*, we are obliged to interpret his claim that G1 destroys the Conservation Program as saying that G1 destroys the Conservation Program *for I*. Thus, it is this claim that we challenged above.

[9] For other statements of the same sentiment see Kreisel [1958], pp. 173–4, and Kreisel [1968], p. 323.

[10] In general, the situation seems to be analogous to that pertaining to the justification of Church's Thesis. Conceived of informally, a body of ideal thought does not possess a mathematically precise description. Therefore, we cannot *mathematically* demonstrate its equivalence to any of its purported (and precisely described) formal explications.

[11] We wish to note expressly that neither Kreisel nor Prawitz explicitly advocates the "unreasonable" formulation of Hilbert's Program just discussed. In a passage not quoted, Kreisel says that the Hilbertian's judgement of the adequacy of a formal system *I* as a codification of some body of ideal reasoning depends upon the *empirical* judgement that the *concepts* of ideal practice can all be expressed in the language of *I* (cf. Kreisel [1968], p. 323). And Prawitz appears to go along with this. Thus, both Kreisel and Prawitz seem to be clearly aware of the need for non-mathematical evidence in judging whether a given formalism is an adequate codification of a given body of informal ideal reasoning. And in addition to this, they are willing to grant that Hilbert himself saw the need for such evidence (which several of his remarks would seem to indicate is true).

Still, they both seem to think that significant harm is done to Hilbert's Program by G1, and it is this that we find perplexing. For, in our opinion, the most that is "shown" by G1 is that one cannot be *mathematically certain* that a given formalism captures all of the instrumentally useful proofs available in a given area of ideal mathematics. [The argument for this application of G1 would, presumably, proceed as follows. It is mathematically possible that some proof of φ (or of $\sim \varphi$) should come to be seen as possessing mathematical utility, and if that were to happen, I (= the ideal system for which φ is the Gödel sentence) would be superceded by $I + \varphi$ (or $I + \sim \varphi$) as the focus of the Hilbertian's attention. Thus, G1 "shows" that for any ideal system I (to which G1 applies) we can never have an *a priori* guarantee that it encompasses all of the useful ideal arguments formulable in its language.]

What we don't see is why the Hilbertian should be bothered by this. Why shouldn't a *non-mathematical* guarantee of the comprehensiveness of his identified ideal methods be enough for him? It seems to us (and we shall presently argue the point) that it would. Thus, we don't understand how Kreisel and Prawitz can, on the one hand, both grant that the Hilbertian neither needs nor claims a mathematical guarantee for the exhaustiveness of his ideal formalisms and, on the other, maintain that G1 does serious damage to his (i.e., the Hilbertian's) program. In other words, given that Kreisel and Prawitz do not appear to advocate the "unreasonable" formulation of Hilbert's Program, we do not see why they take G1 as having such far-reaching significance for it. We shall develop this line of thought in greater detail as our discussion continues.

[12] We would say that one system of ideal methods supercedes another when the epistemic utility of the one exceeds that of the other. See Chapter II for a more detailed description of what the epistemic utility of an ideal system consists of.

[13] In our discussion of the first reading of Smorynski's argument, we pointed up the need for the claim that there is some ideal system proving φ to whose defense the Hilbertian is committed. For all intents and purposes, that claim is the same as the one that we are disagreeing with now. This is so because the only way in which the Hilbertian could come to be committed to defending $I + \varphi$ (i.e., the ideal system formed by adding a formalized version of the intuitive proof of φ to I) is if its instrumental epistemic utility were greater than that of I. And that is what we are now calling into question. It should thus be clear that it is no part of our argument to deny that the Conservation Program fails for $I + \varphi$. φ's being a real formula provable in $I + \varphi$ but not in R is, in our opinion, sufficient to establish that.

However, this fact has serious consequences for the Hilbertian only if his vindicatory program for classical ideal mathematics demands a defense of $I + \varphi$, and our position is that this is not the case. Thus, in essence, what we are saying is that the Hilbertian doesn't need to carry out the Conservation Program for $I + \varphi$, and therefore his inability to do so is a matter of little consequence to him.

[14] These factors are neither mutually exclusive (witness the "overlap" between (1) and (3) due to the fact that the importance of a real product can be related to its extensiveness), nor mutually exhaustive (since another factor affecting the instrumental value of an ideal theory is the strength and feasibility of its soundness proof).

[15] The recent discovery by Paris and Harrington [1977] of a mathematically interesting example of incompleteness in PA might bear thinking about in this connection. They show, of a certain mathematically interesting statement S, that $\vdash_{PA} S \supset Con(PA)$, and hence that not-$\vdash_{PA} S$. Of course, one can also prove that $\vdash_{PA} Con(PA) \supset \varphi$, and hence that $\vdash_{PA} S \supset \varphi$.

What this shows is that for certain choices of I (viz., where I is PA) some extensions of I which entail φ *are* mathematically more interesting than I itself. It does not follow from this, however, that there is any choice of I (PA included) for which $I + \varphi$ is mathematically more interesting than I. (For what Paris and Harrington have shown is that PA + S entails φ; not that PA + φ entails S.) Moreover (and this seems even more significant), there is clearly no reason to suppose that what Paris and Harrington have done for PA (viz., find a mathematically interesting statement formulable but not provable in it) can be done for I (= an arbitrarily selected theory of ideal mathematics) generally. Thus, the most that the Hilbertian would need to do in order to regain stability

in the wake of the Paris—Harrington theorem is to prove the real-soundness of
PA + S. But he is under no obligation to grant the mathematical superiority of
$I + \varphi$ to I generally.

[16] Let us recall once more, however, that we do *not* reject Prawitz' claim that
G1 (and G2 also, for that matter) *do* show that both Hilbertian and anti-Hilbertian alike must pay greater attention to the formal system in which informal
ideal practice is to be codified. As we have already stressed, this *is* a significant
consequence of Gödel's Theorems. But it is vastly weaker than their usual interpretation, and in no way sanctions such an extreme conclusion as that the Hilbert Program cannot be carried out.

[17] And we might add that even if it did, the failure of Hilbert's Program would
not follow since the Hilbertian instrumentalist may not be committed to defending I but rather only certain of its subsystems.

REFERENCES

Ackermann, W.: 1940, 'Zur Widerspruchsfreiheit der Zahlentheorie', *Mathematische Annalen* **117**, 162—194.

Benacerraf, P.: 1973, 'Mathematical Truth', *Journal of Philosophy* **70**, 661—679.

Bernays, P.: 1935, 'Hilberts Untersuchungen über die Grundlagen der Arithmetik', in David Hilbert, *Gesammelte Abhandlungen*, Vol. 3, Springer, Berlin, 1935.

Bernays, P.: 1967, 'Hilbert, David', in P. Edwards (ed.), *The Encyclopedia of Philosophy*, Vol. 3, Macmillan Publishing Co. and the Free Press, New York, 1967.

Bernays, P.: 1935a, 'On Platonism in Mathematics', in P. Benacerraf and H. Putnam (eds.), *Philosophy of Mathematics: Selected Readings*, Prentice-Hall, New Jersey, 1964.

Bernays, P.: 1941, 'Sur les questions methodologiques actuelles de la theorie hilbertienne de la demonstration', in F. Gonseth (ed.), *Les entretiens de Zurich sur les fondements et la méthode des sciences mathématiques, 6—9 decembre 1938*, Leemann, Zurich, 1941.

Berry, G.: 1969, 'Logic Without Platonism', in D. Davidson and J. Hintikka (eds.), *Words and Objections: Essays on the Work of W. V. Quine*, D. Reidel Publ. Co., Dordrecht, Holland, 1969.

Bezboruah, A., and J. C. Shepherdson: 1976, 'Gödel's Second Incompleteness Theorem for Q', *Journal of Symbolic Logic* **41**, 503—512.

Brouwer, L. E. J.: 1912, 'Intuitionism and Formalism', in P. Benacerraf and H. Putnam (eds.), *Philosophy of Mathematics: Selected Readings*, Prentice-Hall, Englewood Cliffs, 1964.

Dummett, M.: 1959, 'Wittgenstein's Philosophy of Mathematics', *Philosophical Review* **68**, 324—348.

Dummett, M.: 1973, 'The Philosophical Basis of Intuitionistic Logic', in Dummett (ed.), *Truth and Other Enigmas*, Harvard Univ. Press, Cambridge, 1978.

Dummett, M.: 1975, 'Wang's Paradox', in Dummett (ed.), *Truth and Other Enigmas*, Harvard Univ. Press, Cambridge, 1978.

Feferman, S.: 1960, 'The Arithmetization of Metamathematics in a General Setting', *Fundamenta Mathematicae* **49**, 35—92.

Field, H.: 1980, *Science Without Numbers: A Defense of Nominalism*, Princeton Univ Press, Princeton.

Frege, G.: 1903, 'Über die Grundlagen der Geometrie', in E.-H. Kluge (trans.

179

and ed.), *On the Foundations of Geometry and Formal Theories of Arithmetic*, Yale Univ. Press, New Haven, 1971.

Frege, G.: 1928, 'Compound Thoughts', in E. D. Klempke (ed.), *Essays on Frege*, Univ. of Illinois Press, Urbana, 1968.

Gandy, R.: 1982, 'Limitations to Mathematical Knowledge', in D. van Dalen, D. Lascar, and T. J. Smiley (eds.), *Logic Colloquium '80*, North-Holland Publishing Co., Amsterdam, 1982.

Gentzen, G.: 1936, 'The Consistency of Elementary Number Theory', in M. E. Szabo (trans. and ed.), *The Collected Works of Gerhard Gentzen*, North-Holland Publishing Co., Amsterdam, 1969.

Gentzen, G.: 1938, 'The Present State of Research into the Foundations of Mathematics', in M. E. Szabo (trans. and ed.), *The Collected Works of Gerhard Gentzen*, North-Holland Publishing Co., Amsterdam, 1969.

Giaquinto, M.: 1983, 'Hilbert's Philosophy of Mathematics', *British Journal for the Philosophy of Science* **34**, 119—132.

Gödel, K.: 1931, 'On Formally Undecidable Propositions of *Principia Mathematica* and Related Systems *I*', in J. van Heijenoort (ed.), *From Frege to Gödel: A Source Book in Mathematical Logic, 1879—1931*, Harvard Univ. Press, Cambridge, 1967.

Gödel, K.: 1958, 'On a Hitherto Unexploited Extension of the Finitary Standpoint', trans. by W. Hodges and B. Watson in *Journal of Philosophical Logic* **9**, 133—142.

Goodstein, R. L.: 1957, *Recursive Number Theory*, North-Holland Publishing Co., Amsterdam.

Hart, W.: 1979, *The Epistemology of Abstract Objects, II*, The Aristotelian Society, Supplementary Volume 53, 153—165, Compton Press Ltd., London.

Hilbert, D.: 1922, 'Neubegrundung der Mathematik', in D. Hilbert, *Gesammelte Abhandlungen*, Vol. 3, Springer, Berlin, 1935.

Hilbert, D.: 1925, 'On the Infinite', in J. van Heijenoort (ed.), *From Frege to Gödel*, Harvard Univ. Press, Cambridge, 1967.

Hilbert, D.: 1927, 'The Foundations of Mathematics', in J. van Heijenoort (ed.), *From Frege to Gödel*, Harvard Univ. Press, Cambridge, 1967.

Hilbert, D.: 1930, 'Naturerkennen und Logik', in D. Hilbert, *Gesammelte Abhandlungen* Vol. 3, Springer, Berlin, 1935.

Hilbert, D., and P. Bernays: 1934—39, *Grundlagen der Mathematik*, Vols. 1—2, Springer, Berlin.

Jeroslow, R.: 1973, 'Redundancies in the Hilbert-Bernays Derivability Conditions for Gödel's Second Incompleteness Theorem', *Journal of Symbolic Logic* **38**, 359—367.

Jeroslow, R.: 1976, 'Consistency Statements in Formal Theories', *Fundamenta Mathematicae* **72**, 2—40.

Kitcher, P.: 1976, 'Hilbert's Epistemology', *Philosophy of Science* **43**, 99—115.

Kneebone, G.: 1963, *Mathematical Logic and the Foundations of Mathematics*, D. van Nostrand, London.

Kreisel, G.: 1958, 'Hilbert's Programme', in P. Benacerraf and H. Putnam (eds.), *Philosophy of Mathematics: Selected Readings*, Prentice-Hall, Englewood Cliffs, 1964.

Kreisel, G.: 1958a, 'Ordinal Logics and the Characterization of Informal Concepts of Proof', in *Proceedings of the International Congress of Mathematicians*, Edinburgh, 1958.

Kreisel, G.: 1958–9, 'Review of Wittgenstein's *Remarks on the Foundations of Mathematics*', *British Journal for the Philosophy of Science* 9, 135–158.

Kreisel, G.: 1965, 'Mathematical Logic', in T. L. Saaty (ed.), *Lectures in Modern Mathematics*, vol. 3, Wiley, New York, 1963/1965.

Kreisel, G.: 1968, 'A Survey of Proof Theory', *Journal of Symbolic Logic* 33, 321–388.

Kreisel, G.: 1970, 'The Formalist-Positivist Doctrine of Mathematical Precision in the Light of Experience', *L'age de la Science* 3, 17–46.

Kreisel, G.: 1971, 'A Survey of Proof Theory II', in J. E. Fenstad (ed.), *Proceedings of the Second Scandinavian Logic Symposium*, North-Holland Publishing Co., Amsterdam.

Kreisel, G.: 1976, 'What Have We Learnt from Hilbert's Second Problem', *AMS Proceedings of Symposia in Pure Mathematics*, vol. 28, American Mathematical Society, Providence.

Kreisel, G., and J.-L. Krivine: 1967, *Elements of Mathematical Logic*, North-Holland Publishing Co., Amsterdam.

Kreisel, G., and A. Levy: 1968, 'Reflection Principles and their use for Establishing the Complexity of Axiomatic Systems', *Zeitschrift für Mathematische Logik und Grundlagen der Mathematik* 14, 97–142.

Kreisel, G. and G. Takeuti: 1974, 'Formally Self-Referential Propositions for Cut-Free Classical Analysis and Related Systems', *Dissertationes Mathematicae* 118, 4–50.

Lear, J.: 1982, 'Aristotle's Philosophy of Mathematics', *Philosophical Review* 91, 161–192.

Leisenring, A. C.: 1969, *Mathematical Logic and Hilbert's ε-Symbol*, Macdonald Technical and Scientific Publishers, London.

Löb, M.: 1955, 'Solution of a Problem of Leon Henkin', *Journal of Symbolic Logic* 20, 115–118.

Montague, R.: 1957, 'Non-finite Axiomatizability', *Summaries of Talks Presented at the Summer Institute of Symbolic Logic, Cornell University 1957*, Princeton Univ. Press, Princeton, 1960.

Mostowski, A.: 1952, 'On Models of Axiomatic Systems', *Fundamenta Mathematicae* 39, 133–158.

Mostowski, A.: 1966, *Thirty Years of Foundational Studies*, Basil Blackwell, Oxford.

Neumann, J. von: 1931, 'The Formalist Foundations of Mathematics', in P. Benacerraf and H. Putnam (eds.), *Philosophy of Mathematics: Selected Readings*, Prentice-Hall, Englewood Cliffs, 1964.

Paris, J., and L. Harrington: 1977, 'A Mathematical Incompleteness in Peano Arithmetic', in J. Barwise (ed.), *Handbook of Mathematical Logic*, North-Holland Publishing Co., Amsterdam.

Parsons, C.: 1967, 'Mathematics, Foundations of', in P. Edwards (ed.), *The Encyclopedia of Philosophy*, vol. 5, Macmillan Publishing Co. and the Free Press, New York, 1967.

Poincaré, J. H.: 1908, *Science and Method,* Translated by F. Maitland, Dover Publications, New York.

Prawitz, D.: 1971, 'Ideas and Results in Proof Theory', in J. E. Fenstad (ed.), *Proceedings of the Second Scandinavian Logic Symposium*, North-Holland Publishing Co., Amsterdam.

Prawitz, D.: 1972, 'The Philosophical Position of Proof Theory', in R. Olson and A. Paul (eds.), *Contemporary Philosophy in Scandinavia*, Baltimore.

Prawitz, D.: 1981, 'Philosophical Aspects of Proof Theory', in *Contemporary Philosophy: A New Survey*, vol. 1, Martinus Nijhoff Publishers, The Hague.

Putnam, H.: 1971, *Philosophy of Logic*, Harper and Row, New York.

Resnik, M. D.: 1974, 'The Philosophical Significance of Consistency Proofs', *Journal of Philosophical Logic* 3, 133–147.

Resnik, M. D.: 1980, *Frege and the Philosophy of Mathematics*, Cornell Univ. Press, Ithaca.

Rosser, J. B.: 1936, 'Extensions of Some Theorems of Gödel and Church', *Journal of Symbolic Logic* 1, 87–91.

Ryll-Nardzewski, C.: 1952, 'The Role of the Axiom of Induction in Elementary Arithmetic', *Fundamenta Mathematicae* 39, 239–263.

Smorynski, C.: 1977, 'The Incompleteness Theorems', in J. Barwise (ed.) *Handbook of Mathematical Logic*, North-Holland Publishing Co., Amsterdam.

Steiner, M.: 1975, *Mathematical Knowledge*, Cornell Univ. Press, Ithaca.

Tait, W.: 1981, 'Finitism', *Journal of Philosophy* 78, 524–546.

Takeuti, G.: 1975, *Proof Theory*, North-Holland Publishing Co., Amsterdam.

Takeuti, G.: 1978, *Two Applications of Logic to Mathematics*, Princeton Univ. Press, Princeton.

Wang, H.: 1970, *Logic, Computers and Sets*, Chelsea Publishing Co., New York.

Webb, J. C.: 1980, *Mechanism, Mentalism, and Metamathematics*, D. Reidel, Dordrecht.

Weyl, H.: 1927, 'Comments on Hilbert's Second Lecture on the Foundations of Mathematics', in J. van Heijenoort (ed.), *From Frege to Gödel*, Harvard Univ. Press, Cambridge, 1967.

Wright, C.: 1982, 'Strict Finitism', *Synthese* 51, 203–282.

INDEX

Ackermann, W. 86, 179
arithmetization 59, 97—101, 103, 104, 105

Benacerraf, P. 41, 179
Bernays, P. 70, 73, 91, 179
  Hilbert and 40, 48, 49, 74, 94, 100, 102, 103, 114, 130, 135
Berry, G. 25, 42, 179
Bezboruah, A. (and J. C. Shepherdson) 135, 158, 179
Boolos, G. 131
Brouwer, L. E. J. xi, xii, 14, 75, 179
Byrd, M. 131

Consistency Program 161, 174, 175, 176
contentual reasoning (see also finitary reasoning)
  vs. non-contentual reasoning 3—9, 12, 36, 39—40, 49—53
  mathematical induction conceived as 59—62
  problematic vs. unproblematic varieties of 18—19, 39, 40, 46, 47, 70, 74, 90, 92, 154
Conservation Program 161, 164, 165, 166, 169, 173, 174, 175, 176, 177
conservativeness (see also real-soundness) 31, 41
Convergence Problem (for the Standard Argument) 77, 78, 85—86, 90, 142, 143, 144, 147, 152—155, 157, 159

Derivability Conditions 94—97, 130
  Mostowskian defense of 101—113
  Kreisel-Takeuti defense of 113—124, 132, 136, 137, 138
  Classical (?) defense of 124—129
Diagonalization Property (DIAG) 94, 95, 96, 131
  formalized (F-DIAG) 96
Dilution Problem (for the Standard Argument) xiii, 1, 17, 21, 22, 35, 36, 39, 41, 42, 43, 44, 46, 62—73, 75, 90, 92, 129, 175
  and Hilbertian finitism 35—36, 40—41, 43—44, 62—73
Dummett, M. A. E. 36, 76, 157, 179

epsilon-elimination strategy 7, 92
extrinsic revision 122—124, 138, 139, 146

Feferman, S. 91, 100, 102, 103, 131, 138, 179
Field, H. 1, 2, 22—26, 41, 42
  nominalism of 2, 22—24, 41, 42
finitary reasoning (see also contentual reasoning) 16—22, 42, 46—59
  Cartesian indubitability of 46, 66—71
formalism xiii, 38, 40, 70
formalized modus ponens (F-MP) 96, 115, 130, 131, 133, 135

183

Frege, G. xiii, 1, 9—16, 25, 38, 39, 179, 180
Frege's Problem (for instrumentalism) xiii, 1, 9—16, 21—26, 35, 36, 38, 42, 45, 57, 59, 74, 84, 91, 175

Gandy, R. O. 157, 180
Gentzen, G. 36, 37, 73, 82, 86, 180
Giaquinto, M. 139, 180
Gödel, K. 50, 51, 53, 73, 78, 81, 91, 94, 101, 145
  first incompleteness theorem of (G1) xii, xiii, 94—95, 104, 160—177
  second incompleteness theorem of (G2) xii, xiii, 36, 73, 77, 78, 82, 86, 89, 91, 94, 95, 97, 100, 101, 102, 104, 113, 114, 128, 129, 130, 131, 140, 144, 146, 147, 152, 155, 157, 158, 159, 160, 161, 164, 174, 175, 176, 178
Gödelian Challenge (to Hilbertian instrumentalism) 77—92, 147, 160
Goodstein, R. L. 74, 180

Harrington, L. (and J. Paris) 177—178, 181
Hart, W. 24—25, 42, 180
Henkin, L. 130
Hilbertian instrumentalism xii, 1, 2, 3—24
  vs. neo-Fregean realism 24—34
  Darwinistic argument for 8—9, 37
  Equivalency Thesis of 20—23, 42
Hilbertian residues 89, 90, 144, 149, 150, 151, 152, 153, 156
Husserl, E. 38
hypothetical judgements 40, 47, 73

ideal reasoning 3—21, 37—38
  vs. real reasoning (see contentual vs. non-contentual reasoning)
  mathematical induction as a form of 14—15, 59—62
Indispensability Thesis (Quine's) 24
induction (mathematical) 15, 45, 46, 60, 61, 62
  the status of 46—59
infallibilist argument 66, 71
instrumental utility
  and efficiency 29—30, 32, 43
  and acuity 29—30
    perspicacity 29—31
    reliability 29—30, 31—32, 43
interpretation (of a formalism)
  vs. evaluation 12—14, 34—36, 43
intrinsic revision 123
intuitionism xii, 38, 50, 73, 74, 76

Jeroslow, R. 100, 130, 131, 158, 180

Kantian intuition 64—66
Kitcher, P. 36, 46, 64, 65, 66, 180
Kneebone, G. T. 92, 180
Kreisel, G. 36, 40, 52, 53, 70, 71, 73, 74, 76, 91, 113, 135, 136, 139, 166, 167, 169, 176, 180, 181
  and J.-L. Krivine 53, 71, 181
  and A. Levy 91, 136, 181
  and G. Takeuti 91, 130, 134, 138, 181
Kreisel-Takeuti Proposal (concerning the defense of the Derivability Conditions) 113—124, 132, 136, 137, 138
Kreisel-Prawitz argument (concerning the anti-Hilbertian application of G1) 165—170, 173, 174, 176, 177

Lear, J. 26, 42—43, 181
Leisenring, A. 92, 181
Levy, A. (and G. Kreisel) 91, 181
Löb, M. 100, 130, 181
localization 142, 143, 144, 145—148, 150, 155, 160
  conditional vs. unconditional 145—147, 156, 158, 159
Local Provability Completeness (LPC) 94, 95, 96, 99, 111, 112, 129, 130, 131, 132, 133, 134, 136
  formalized (F-LPC) 96, 101, 111, 112, 114, 115, 125, 128, 130, 131, 134, 135, 136, 138
Local Provability Soundness (LPS) 94, 95, 129

Metamathematical Replacement Strategy (of Hilbertian instrumentalism) 14, 18, 39, 63, 75, 84, 88, 91, 92, 143, 149, 158, 159, 160
Mino, T. 42
Montague, R. 158, 181
Mostowski, A. 82, 101—113, 131, 132, 133, 134, 151, 152, 155, 158
Mostowski's Proposal (concerning the defense of the Derivability Conditions) 101—113
M-proof 57, 63, 71, 72, 74, 75, 84, 85, 87, 88, 90, 92, 149, 157, 158, 159, 160

naturalistic paradigm (for explaining the epistemic utility of a system) 28
Neumann, J. von 7, 33, 36, 37, 181
nominalism (see under Field, H.)

O-proof 57, 63, 71, 72, 74, 75, 87, 88, 90, 92, 149, 158, 159, 160

Paris, J. (see Harrington, L. and J. Paris)
Parsons, C. 72—74, 182
partial propositions 19, 40
Peano Arithmetic (PA) 89, 92, 143, 145, 148, 150, 151, 152, 153, 154, 155, 156, 158, 159, 160, 177
Poincaré, J. xiii, 14—16, 35, 36, 39, 45, 46, 59—62
Poincaré's Claim (concerning the circularity of Hilbertian instrumentalism) 60—61
Poincaré's Problem (for Hilbertian instrumentalism) xiii, 14, 15, 16, 35, 36, 39, 45, 59—62
Prawitz, D. 25, 26, 36, 43, 124, 126, 128, 139, 166, 167, 169, 176, 178, 182
Primitive Recursive Arithmetic (PRA) 45, 58, 59, 74, 75, 140, 154, 155, 159
Principle of Weak Optimality (PWO) 71—73, 90
problematic reals (see under contentual reasoning)
Putnam, H. 23, 41, 182
Π_1^0-reflection 125—126, 139, 140
Π_1^0-soundness 126, 127

Quine, W. V. O. 2, 22
  realism of 24—25, 32, 33, 35

realism (see also under Quine) 2, 3, 24, 25, 27, 35, 42
  neo-Fregean 25—28, 30, 32, 33, 35, 43
real-completeness 126, 127, 128, 164, 175, 176
real-soundness 15, 31, 35, 37, 43, 46, 54, 57, 58, 59, 60, 61, 62, 63, 64, 67, 68, 69, 71, 72, 73, 74, 75, 79, 86, 87, 88, 90, 91, 92, 117, 121,

124, 125, 126, 127, 128, 129, 137, 139, 140, 142, 143, 149, 150, 151, 152, 154, 157, 158, 159, 160, 164, 166, 168, 170, 173, 175, 177, 178

reflexivity 151, 152, 155, 158

reliability (see also real-soundness) 2, 3, 6, 12, 19, 20, 21, 22, 26, 29, 30, 31, 32, 37, 43, 63, 64, 85, 87, 121

replicationist strategy (for explaining the epistemic utility of a system) 28—35

Resnik, M. D. 40, 91, 182

revision (see extrinsic revision and intrinsic revision)

Rosser, J. B. 122, 131, 134, 138, 182

Rosser variant 122—123, 129, 138, 139, 140—141, 145, 147, 156, 159, 160

Russell, B. xii

Ryll-Nardzewski, C. 158, 182

Shepherdson, J. (see Bezboruah, A. and J. Shepherdson)

Smorynski, C. 129, 139, 161, 164, 165, 166, 173, 174, 175, 176, 177, 182

Steiner, M. 75, 182

Stability Problem (for the Standard Argument) 77, 78, 80—83, 90, 93—140

Standard Argument (SA) (against Hilbert's Program) 77, 78—80, 81, 85, 86, 88, 89, 90, 97, 131, 143, 144, 145, 157

Standard Strategy (for arguing against Hilbert's Program) 144, 146, 147, 148, 150, 151, 152, 153, 154, 155, 158, 159

strict finitism 69, 76

strict instrumentalism 83—85

problem of (for the Standard Argument) 77, 78, 86—90, 142, 144, 148—152, 157

thesis of (TSI) 84—85, 86, 89, 91, 142, 143, 150, 153, 157, 158

Σ_1^0-completeness 115, 116, 117, 118, 120, 127, 136, 138

demonstrable 118, 121, 123, 125, 126, 127, 128, 134, 135, 139, 140

Tait, W. 41, 45, 46, 49, 58, 59, 64, 65, 66, 67, 68, 71, 74, 76, 182

Takeuti (see also Kreisel, G. and G. Takeuti) 74, 91, 156, 182

Unavoidability (of finitary reasoning) 66—69, 76

strong local 67, 68—69

weak local 67, 68

strong global 67, 69

weak global 67, 68

Wagner, S. 39

Wang, H. 76, 182

Webb, J. C. 74, 182

Weyl, H. 37, 38, 182

Whitehead, A. N. xii

Wright, C. 76, 182

ZF 89, 92, 143, 145, 155, 158